Common Conversion Factors

Length

1 yard = 3 ft, 1 fathom = 6 ft

	in	ft	mi	cm	m	km
1 inch (in) =	1	0.083	1.58×10^{-5}	2.54	0.0254	2.54×10^{-5}
1 foot (ft) =	12	1	1.89×10^{-4}	30.48	0.3048	
1 mile (mi) =	63,360	5,280	1	160,934	1,609	1.609
1 centimeter (cm) =	0.394	0.0328	6.2×10^{-6}	1	0.01	1.0×10^{-5}
1 meter (m) =	39.37	3.281	6.2×10^{-4}	100	1	0.001
1 kilometer (km) =	39,370	3,281	0.6214	100,000	1,000	1

Area

1 square mi = 640 acres, 1 acre = 43,650 ft^2 = 4046.86 m^2 = 0.4047 ha
1 ha = 10,000 m^2 = 2.471 acres

	in^2	ft^2	mi^2	cm^2	m^2	km^2
1 in^2 =	1		—	6.4516	—	—
1 ft^2 =	144	1	—	929	0.0929	—
1 mi^2 =	—	27,878,400	1	—	—	2.590
1 cm^2 =	0.155	—	—	1	—	—
1 m^2 =	1,550	10.764	—	10,000	1	—
1 km^2 =	—	—	0.3861	—	1,000,000	1

Other Conversion Factors

1 ft^3/sec = .0283 m^3/sec = 7.48 gal/sec = 28.32 liters/sec

1 acre-foot = 43,560 ft^3 = 1,233 m^3 = 325,829 gal

1 m^3/sec = 35.32 ft^3/sec

1 ft^3/sec for one day = 1.98 acre-feet

1 m/sec = 3.6 km/hr = 2.24 mi/hr

1 ft/sec = 0.682 mi/hr = 1.097 km/hr

1 billion gallons per day (bgd) = 3.785 million m^3 per day

1 atmosphere = 1.013×10^5 N/m^2 = approximately 1 bar

1 bar = approx. 10^5 N/m^2 = 10^5 pascal (Pa)

Commonly Used Multiples of 10

Prefix (Symbol)	Amount	Prefix (Symbol)	Amount
exa (E)	10^{18} (million trillion)	centi (c)	10^{-2} (one-hundredth)
peta (P)	10^{15} (thousand trillion)	milli (m)	10^{-3} (one-thousandth)
tera (T)	10^{12} (trillion)	micro (μ)	10^{-6} (one-millionth)
giga (G)	10^{9} (billion)	nano (n)	10^{-9} (one-billionth)
mega (M)	10^{6} (million)	pico (p)	10^{-12} (one-trillionth)
kilo (k)	10^{3} (thousand)		

Volume

	in^3	ft^3	yd^3	m^3	qt	liter	barrel	gal. (U.S.)
1 in^3 =	1	—	—	—	—	0.02	—	—
1 ft^3 =	1,728	1	—	.0283	—	28.3	—	7.480
1 yd^3 =	—	27	1	0.76	—	—	—	—
1 m^3 =	61,020	35.315	1.307	1	—	1,000	—	—
1 quart (qt) =	—	—	—	—	1	0.95	—	0.25
1 liter (l) =	61.02	—	—	—	1.06	1	—	0.2642
1 barrel (oil) =	—	—	—	—	168	159.6	1	42
1 gallon (U.S.) =	231	0.13	—	—	4	3.785	0.02	1

Energy and Power

1 kilowatt-hour = 3,413 Btus = 860,421 calories

1 Btu = 0.000293 kilowatt-hour = 252 calories = 1,055 joule

1 watt = 3.413 Btu/hr = 14.34 calories/min

1 calorie = the amount of heat necessary to raise the temperature of 1 gram (1 cm^3) of water 1 degree Celsius

1 quadrillion Btu = (approximately) 1 exajoule

1 joule = 0.239 calorie = 2.778×10^{-7} kilowatt-hour

Mass and Weight

1 pound = 453.6 grams = 0.4536 kilogram = 16 ounces

1 gram = 0.0353 ounce = 0.0022 pound

1 short ton = 2,000 pounds = 907.2 kilograms

1 long ton = 2,240 pounds = 1,008 kilograms

1 metric ton = 2,205 pounds = 1,000 kilograms

1 kilogram = 2.205 pounds

ENVIRONMENTAL ISSUES

MEASURING, ANALYZING, AND EVALUATING
SECOND EDITION

Robert L. McConnell, Ph.D.
Department of Environmental Science and Geology
Mary Washington College
Fredericksburg, Virginia

Daniel C. Abel, Ph.D.
Department of Marine Science
Coastal Carolina University
Conway, South Carolina

PRENTICE HALL Upper Saddle River, NJ 07458

Library of Congress Cataloging-in-Publication Data

McConnell, Robert L.
    Environmental issues: measuring, analyzing and evaluating / Robert L. McConnell,
    Daniel C. Abel.—2nd ed.
        p. cm.
    Includes bibliographical references.
    ISBN 0-13-092041-X
        1. Environmental sciences—Study and teaching (Higher)   2. Pollution—Study and
    teaching (Higher)   I. Abel, Daniel C.   II. Title.
GE70 M38 2002
372.3'57—dc21   2001026712

Executive Editor: *Dan Kaveney*
Assistant Editor: *Amanda Griffith*
Editorial Assistant: *Margaret Ziegler*
Production Editor/ Composition: *Pine Tree Composition*
Executive Managing Editor: *Kathleen Schiaparelli*
Assistant Managing Editor: *Beth Sturla*
Marketing Manager: *Christine Henry*
Managing Editor, Audio/Video Assets: *Grace Hazeldine*
Art Editor: *Shannon Sims*
Art Director: *Jayne Conte*
Cover Designer: *Bruce Kenselaar*
Photo Researcher: *Diana P. Gongora*
Manufacturing Manager: *Trudy Pisciotti*
Assistant Manufacturing Manager: *Michael Bell*
Vice President of Production and Manufacturing: *David W. Riccardi*

 2002, 1999 by Prentice-Hall, Inc
Upper Saddle River, New Jersey 07458

Printed in the United States of America

10 9 8 7 6 5 4 3

ISBN 0-13-092041-X

Pearson Education Ltd., *London*
Pearson Education Australia Pty., Limited, *Sydney*
Pearson Education Singapore, Pte. Ltd.
Pearson Education North Asia Ltd., *Hong Kong*
Pearson Education Canada, Ltd., *Toronto*
Pearson Educación de Mexico, S.A. de C.V.
Pearson Education—Japan, *Tokyo*
Pearson Education Malaysia, Pte. Ltd.

For Brian
RLM

For Mary, Juliana, & Louis
DCA

CONTENTS

Part Five ☺ The Hydrosphere

Part Six ☺ Environmental Health

Preface

To the Instructor

The idea for this book arose when we were colleagues at Mary Washington College. We grew impatient with a teaching style centered on the faculty member as "lecturer" and expert and the student as "scribe" and novice. We feel that such an approach encourages students to be passive rather than active learners and leads to an unhealthy dependency on the faculty person as "expert."

Although students generally are both capable and dedicated, many are also afraid of math, rusty in its use, or were superficially trained in arcane fields of calculus. This lack of math skills often leaves students unprepared to deal with the complexity of today's environmental issues.

Moreover, we are continually surprised to discover how many bright students can't do three things: understand and confidently manipulate the units of the metric system, use scientific notation, or critically evaluate complex environmental issues.

Most of a student's discomfort with math is generally founded in frustration. For example, making one error in a series of calculations can render the whole effort useless. We believe that, in the absence of a real learning disability, to solve most math problems requires no special aptitude, only clear sequential instructions, attention to detail, and hard work. That is why step-by-step calculations are included in your Answer Key.

One of our major objectives is thus to help develop math literacy (*numeracy*) among today's students. We understand that many students have some "math anxiety," so we have included in this second edition a section entitled *Using Math in Environmental Issues,* in which we use a step-by-step method to take students through examples of the calculations in the Issues. We even show sample keystrokes involved in using common calculators. We believe this method will gradually build the student's confidence enough to trust in his or her own efforts. Math proficiency is one of the important skills necessary for fully understanding environmental issues, and without these skills, the student's only option is to make choices on the basis of which "expert" is most "believable." Such skills involve the ability to manipulate large numbers using scientific notation and exponents, the ability to use compound growth equations containing natural logs, and so on.

It is our goal that this book be provocative, factually accurate and up-to-date. *Environmental Issues: Measuring, Analyzing, and Evaluating* is meant to be the basis for an issues-oriented introductory, seminar, upper-level, or laboratory course in environmental science or studies. However, it can also be used as a supplement to traditional texts in environmental science, geology, biology, and other natural sciences and in humanities courses that seek to cultivate an awareness of and knowledge about environmental issues (*ecolacy*). These issues may be viewed as projects that use pressing environmental issues as a means to develop students' critical thinking skills in a deliberate and structured way. By their nature, they require students to integrate topics from across subdisciplines to measure, analyze, and evaluate each issue using the discipline and method of a scientist.

But becoming educated is much more than simply acquiring skills. Therefore, we have two additional objectives: to provide students with the knowledge and intellectual standards necessary to apply critical thinking to environmental studies and to foster their ability to critically evaluate issues.

As such, *Environmental Issues: Measuring, Analyzing, and Evaluating* is as much an **interactive** workbook as a traditional textbook. We expect students to have access to standard references in environmental, physical, and natural sciences and to have access to and know how to use the World Wide Web. Indeed, every Issue contains URLs (Uniform Resource Locators) to websites for up-to-the-minute information.

We also trust that you, as an expert in your field and with your own perspectives, will supplement the information in this book with comments, introductions, and your own critical comments on the questions we ask your students to consider. The questions in each issue have been used by students in a university/college setting. The issues have undergone an exhaustive set of permutations to make them as user-friendly and comprehensible as we can make them. The issues are based on solid science, key terms are defined, and jargon is kept to a minimum. When important terms are introduced they are italicized and defined if necessary. Key mathematical formulas are introduced and painstakingly explained using a step-by-step, nonthreatening approach that we believe you will appreciate.

But we designed this book to be more than a workbook. As we mentioned, one of our major objectives is to foster numeracy among today's students—not necessarily arcane math, but the kind of math needed to properly quantify environmental issues, such as the use of key formulas, scientific notation, and the metric system. We provide detailed introductions for each of these topics, as well as a detailed Answer Key for you to use as you see fit that shows the step-by-step calculations used to determine the answers. We have purposefully not directly provided students access to the Answer Key. In addition, we provide you with suggested answers to the "Critical Thinking" questions, but we are confident that you will have your own point of view that you will wish to develop in many if not most of them.

To encourage rigorous critical thinking, each issue has a set of "Critical Thinking" questions with spaces for answers. Critical thinking involves using a set of criteria and standards by which the reasoner constantly assesses her/his thinking. At the core of critical thinking is self-assessment.

These questions are intended to serve two purposes: to allow students to practice their writing in a non-threatening journal format and to develop their critical thinking skills. In terms of critical thinking, we devote a detailed section to the aspects of this important concept in the first section of this book. *It is vitally important that students read this material for content as they will be asked to apply these standards and criteria throughout the book.*

We trust that this book will provide an extraordinarily stimulating experience for your students. The "Critical Thinking" questions have been designed (using principles of critical thinking) to be provocative, and you or your students may perceive a "bias" in the wording of some of the questions. Although we have made the content as factual as possible, we do have strong convictions about these issues. *Convictions are not, however, biases.* Based on the scientific method, our views as scientists are subject to change as evidence supporting our convictions changes. And as such, we are constantly testing the *assumptions* that we use when approaching complex environmental issues. Indeed, this aspect can be turned to a major advantage. Ask your students to look for examples of bias in the questions, and then discuss with them the difference in science between "bias" and "conviction." No doubt it will prove a fruitful activity and may lead students into research (perhaps to "prove us wrong"), which is the essence of progress in the search for scientific truth.

How to Use This Book Effectively

As an instructor, you are probably aware that many students will resist doing the work necessary to fully appreciate the issues. Others will wait until the last minute and then slap together something just in time to turn in. Still others will respond to critical thinking questions with only a few words containing inch-high letters.

Here's how we use *Environmental Issues* in our classes. First, we typically dedicate a class or a portion of a laboratory early in the semester to introducing students to the principles of critical thinking. In this period we ask students to list characteristics of critical thinking (or higher order thinking, or just plain good, effective thinking). More often than not, the class's list encompasses many of the standards of critical thinking that are contained in the section of this book entitled *Basic Concepts and Tools: Using Math and Critical Thinking.* As we go over these characteristics, we emphasize *clarity, awareness of assumptions,* and *continuous self-assessment,* as well as the importance of applying critical thinking to environmental (and other) issues. We then analyze passages from letters to the editor, newspaper op-eds, and popular magazine articles.

It is also worthwhile, before using this book, to confront students' math anxiety early and attempt to reassure them that they are capable of doing the math, although they may be rusty and require some practice and assistance.

We frequently assign Issues or blocks of Issues as group projects, in which students collaborate using a web page, "blackboard" or another form of electronic discussion group, and we try to ensure that every student logs on to the system. An e-mail system allows your students to communicate with each other outside of class; they can exchange information, send reports, references, and critiques to classmates, and so on. Using this format, students can also provide less-threatening critiques of each other's work and ideas. We recommend you encourage students to do this, while reminding them that few people are won over to another's argument by having their own ideas ridiculed.

For classes of more than 24 students, use of a group e-mail system offers you enormous possibilities. You can break the class into working groups of three to four members, and the group can be reorganized as the term progresses. Working groups can do the calculations independently, can check their work by e-mail, or in meetings, and can get together to hash out the answers to the *For Further Study* questions.

This method also allows you to monitor the class's work and communicate with the class in a nonthreatening manner. Your institution's computer or network services department can provide you with the details if you've never used one.

We have found that many students who would be reluctant to participate in a classroom discussion will willingly contribute in the relative anonymity of an electronic discussion group. Students can (and probably should) copy their math and send it to others over the system, thereby checking each other's work. You can send comments to the groups if you feel they are on the wrong track and commend them and encourage them if they are making progress. You can send them questions arising out of their own discussions and respond immediately to their inquiries. They can exchange information on the *For Further Study* questions, and you can discuss these question with the students as their work evolves.

Regardless of whether you are teaching a large or small class, you can have students work in groups of two to four members. They can debate the issues and grade each other, or they can turn in their work in the normal fashion and be graded on the accuracy of their calculations as well as on the thoroughness of their answers to selected *For Further Study* questions.

Whatever the size of your class, work on these issues can take the place of one or more exams, freeing you up for other activities and providing students with a less-threatening way to earn class credit than an all-exam format. We believe that students retain more information from work on projects and reports than from cramming for tests. At the end of the term, you could pass out a study sheet detailing specifically what material from the issues will be covered on the final exam, if you choose to test them on their work in this manner.

A Few Final Comments

This book has a simple, straightforward format and intentionally includes only black and white figures. We made this decision to lower printing costs and thus save your students money. We realize, however, the importance of illustrations: As our editors constantly remind us, texts must be able to pass the "flip test."

For location maps not included in the text, we recommend that students access Census Bureau data at http://www.census.gov and Tiger Map Server Browser at http://www. nwbuildnet.com/nwbn/mapservice.html, which is an outstanding website for maps. At this site students can type in city, county, or state names or use a zip code to access maps and demographic information for any locality in the United States. These maps can also be manipulated to provide representation of additional data and can be downloaded and printed.

This book provides you with a wide range of issues, many of which you and your students may never have thought about previously. It is likely that you will not be able to get through all of these in one semester, unless you assign different issues to groups of students. However, we feel Issue 1 is pivotal, and it should be completed first and very carefully. Other issues can be assigned at your discretion.

You may contact us personally at rmcconne@mwc.edu and dabel@coastal.edu with questions and comments. We hope to hear from you!

TO THE STUDENT

If you are concerned about and interested in environmental issues, but feel you just "can't deal with the math," then this book is for you. As environmental scientists, we care deeply about environmental issues, and we feel that you, as a responsible citizen who will have to make increasingly difficult choices in the years ahead, need to be concerned about them as well. Here are a few of our reasons:

- As this book goes to press, the world's leaders are beginning a dialogue about how to respond to human-induced global climate change, a result of global warming. A scientific consensus already exists that our growing human population is having a measurable effect on the composition of the planet's atmosphere. Even though we can't yet be certain what the effect of these changes will be, the preponderance of evidence suggests the impact will, on the whole, be negative, and could be catastrophic for hundreds of millions of people crowded into many of the planet's coastal cities.
- Oceanographers are becoming increasingly concerned about the very survival of many marine organisms, including some that form the basis of major world fisheries. In fact, entire ecosystems, such as coral reefs, may be at risk. The ocean's ability to absorb our waste, toxic and otherwise, is certainly limited, and these limits are being tested.
- Although we in the United States and in a few other areas of the world have made impressive strides in improving or at least slowing the degradation of our air, water, and soil, our relentlessly growing human numbers threaten this progress.
- Growing levels of material consumption in developed countries, coupled with less-regulated international commerce (free trade) are placing increased stress on critical ecosystems such as tropical hardwood forests, which in turn gravely threatens the planet's species diversity.

- Environmental issues, such as water conflicts in the Middle East and the impact of air and water pollution, could destabilize international relations, thus leading to regional conflicts involving powers such as the United States.
- Fossil fuels are the basis of industrial and postindustrial society. Their extraction, transportation, and use impose significant costs on the planet. Much of this cost is externalized (or dumped) onto the environment. In our view, addressing environmental issues must involve an awareness of these costs.

We hope you will find this book to be a provocative introduction to a number of these issues and to many others that you may have never even thought about. These are real-life issues, not hypothetical ones, and you need certain basic skills to fully understand them.

- You must be familiar with the units of the metric system.
- You must be able to use a few mathematical formulas to quantify the issues you will be debating, and you must be able to carry out the calculations accurately.
- You should develop the habit of rigorously assessing your thinking and you should apply certain critical thinking skills and techniques when discussing the implications of your calculations.
- You should realize that critical thinking requires practice, dedication, and time.

We understand that many students have some "math anxiety," so we use a step-by-step method to take you through many of the calculations in this book. We have even included samples of keystrokes used on common hand-held calculators. Math proficiency is one of the important skills necessary for fully understanding environmental issues, and without these skills, your only option is to make choices on the basis of which "expert" you believe.

But becoming educated is much more than simply acquiring skills. Therefore, we have another two fundamental objectives: to provide you with the knowledge and intellectual standards necessary to apply critical thinking to environmental studies and to foster your ability to critically evaluate issues, without which mechanical skills are of limited value. To that end, we have included a section, *Basic Concepts and Tools: Using Math and Critical Thinking,* in which we illustrate and describe each criterion for critical thinking, as well as the framework within which these criteria are applied. It is very important that you read this section carefully and do all of the exercises in it *before* you begin the analyses of the issues.

Let's consider an example: *growth.* The word is used in many societal, economic, demographic, and environmental contexts, including growth of the economy, growth of the population, growth of impervious surfaces, growth of food production, and growth of energy use. Assessing the impact of growth requires an understanding of a few simple equations such as the compound interest equation. You must be able to use it accurately and understand its implications.

As our national and global population grows and changes and our relationships with other nations and peoples evolve, environmental issues will become increasingly complicated. Domestically, demographic and ethnic changes are becoming more important, which in turn requires an enhanced ability to critically evaluate issues.

We don't try to avoid controversial topics such as population growth, personal consumption, the automobile, and immigration. We hope you will be challenged by the issues discussed in this text and that you will research them and become an "expert" on the topics yourself. In fact, if we may be allowed a hidden agenda, this is it.

And while you are testing your own thinking and reasoning, test ours as well. Look for examples of "bias" in the "Critical Thinking" questions and be prepared to explain your conclusions and discuss them with your fellow students and with your instructor.

A Note on Conventions Used in This Book

When we express rates, concentrations, and so on—for example, milligrams per liter, people per hectare, tonnes per year—we use all acceptable formats, including "per" (as above), the slash (mg/L, people/hectare, tonnes/year), or at times negative exponents ($mg \cdot L^{-1}$, people $\cdot$ hectare^{-1}, tonnes $\cdot$ year^{-1}). These have been used interchangeably throughout the book.

Information sources are provided as footnotes at the bottom of each page and the corresponding number is placed in the text where the footnote is cited.

ACKNOWLEDGMENTS

This book could not have succeeded without the aid of numerous friends and colleagues. We appreciate the invaluable feedback from the following reviewers, whose comments and suggestions improved the manuscript exponentially.

Marvin Baker, *University of Oklahoma*
George Bean, *University of Maryland, College Park*
Grady Blount, *Texas A & M University, Corpus Christi*
Robert W. Bowker, *Glendale Community College*
John A. Bumpus, *University of Northern Iowa*
Carl Chuey, *Youngstown State University*
James Dyer, *Ohio University*
Barry C. Field, *University of Massachusetts, Amherst*
Elizabeth Forys, *Eckerd College*
Christopher Fox, *Catonsville Community College*
Cheryl Greengrove, *University of Washington*
Bill Hahn, *Columbia University*
Robert Hilton, *Massasoit Community College*
George Hinman, *Washington State University*
Tim Holtsford, *University of Missouri, Columbia*
Alan R. Holyoak, *Manchester College*
Barbara Holzman, *San Francisco State University*
Jeffrey Jack, *Western Kentucky University*
David J. Johnson, *Michigan State University*
Erik Lindquist, *Lee University*
Matthew McConeghy, *Johnson & Wales University*
Steve Obenauf, *Broward Community College*
Nancy Ostiguy, *California State University, Sacramento*
Clayton Penniman, *Central Connecticut State University*
Steven C. Steel, *Bowling Green State University*
Ronald C. Sundell, *Northern Michigan University*
Jack A. Turner, *University of South Carolina, Spartanburg*
Maud Walsh, *Louisiana State University*
D. Reid Wiseman, *The College of Charleston*

We also thank Daniel Kaveney, Beth Sturla, and their staff at Prentice Hall and Patty Donovan and the staff at Pine Tree Composition. Periodically throughout the writing of this work, our think tank and refuge was Sonoma State University's Center for Critical Thinking, where this book began to take shape during its annual conference.

Robert L. McConnell

Daniel C. Abel

Rather than to travel into the sticky abyss of statistics,
it is better to rely on a few data and on the pristine
simplicity of elementary mathematics.

—ALBERT BARTLETT

BASIC CONCEPTS AND TOOLS: USING MATH AND CRITICAL THINKING

In this introductory chapter we will review the units of the metric system and the rules for using scientific notation in math problems, which you will need in order to analyze the issues in this book. We will also show you how to do slightly more complicated math, such as projecting population based on growth rates, using step-by-step methods. This chapter also deals with another basic skill you need to practice quantitative environmental analysis: *critical thinking*. At first glance many issues seem perplexing and hard to approach. Never fear; we will show you how to apply intellectual standards within a critical thinking framework.

THE METRIC SYSTEM

The metric system's elegance and utility arise from its simplicity. You probably already know that the metric system is based on powers of ten: 10 millimeters in 1 centimeter, 100 centimeters in 1 meter, and so forth. This key point should be kept in mind. Once you understand this system you will never be satisfied with the "English" system again; you will undergo an epiphany.

You must be able to make certain important conversions from the English system to the metric system and back. (It is a good idea to mark this page, since you will find it useful to refer to again.) Some common conversions are given in Table 1 below (and also inside the front cover of this book).

One category of conversions involves *area*. Use the conversion factors in Table 1-2.

Table 1 ■ Some Common Metric Conversions

gallons/liters	1 U.S. gal. = 3.8 L	One U.S. gallon = 3.8 liters
liters/gallons	1 L = 0.264 U.S. gal.	One liter = 0.264 U.S. gallons
meters/yards	1 m = 1.094 yd	One meter = 1.094 yards
yards/meters	1 yd = 0.914 m	One yard = 0.914 meters
grams/ounces	1 g = 0.035 oz	One gram = 0.035 ounce
ounces/grams	1 oz = 28.35 g	One ounce = 28.35 grams
kilograms/pounds	1 kg = 2.2 lb	One kilogram = 2.2 pounds
pounds/grams	1 lb = 454 g	One pound = 454 grams
miles/kilometers	1 mi = 1.609 km	One mile = 1.609 kilometers
kilometers/miles	1 km = 0.621 mi	One kilometer = 0.621 mile

Table 2 ■ Conversion Factors for Area

square miles/square kilometers	1 mi² = 2.6 km²	One square mile = 2.6 square kilometers
square kilometers/square miles	1 km² = 0.39 mi²	One square kilometer = 0.39 square miles
hectares/acres	1 ha = 2.47 acres	One hectare = 2.47 acres
acres/hectares	1 acre = 0.4 ha	One acre = 0.4 hectares
square yards/square meters	1 yd² = 0.84 m²	One square yard = 0.84 square meters
square meters/square yards	1 m² = 1.2 yd²	One square meter = 1.2 square yards

Here's a useful shortcut. To convert areas, you can use the conversion factors for the units of length and square them. For example, to convert 6 square feet to square yards, do the following:

$$6 \text{ ft}^2 \times (1 \text{ yd}/3 \text{ ft})^2 = 6 \cancel{\text{ft}}^2 \times 1 \text{ yd}^2/9 \cancel{\text{ft}}^2 = 0.67 \text{ yd}^2$$

Another frequently used conversion within the metric system is that for water: *There are 1000 liters per cubic meter.*

The important units of the metric system used in this book, along with some of their equivalents, are given in Table 3. There are other units of measure in the metric system, and we will introduce and explain them as the need arises.

Now do these self-assessment questions to test your ability to manipulate and convert these units. These skills are essential for the work in the issues that follow. Remember to use your conversion factors (so that units cancel each other out). For example, to determine how many liters are in 3.6 cubic meters:

$$3.6 \cancel{\text{m}}^3 \times 1000 \text{ L}/\cancel{\text{m}}^3 = 3600 \text{ L}$$

Table 3 ■ Units of the Metric System: Conversions

Units of Distance: The fundamental unit is the *meter*	
1000 (10^3) m = 1 km	One thousand meters = one kilometer
100 (10^2) cm = 1 m	One hundred centimeters = one meter
10 (10^1) mm = 1 cm	Ten millimeters = one centimeter
1000 (10^3) μm = 1 mm	One thousand micrometers = one millimeter
Units of Mass: The fundamental unit is the *gram*	
1000 kg = 1 metric ton	One thousand kilograms = one metric tonne (spelled tonne)
1000 g = 1 kg	One thousand grams = one kilogram
1000 mg = 1 g	One thousand milligrams = one gram
1,000,000 (10^6) ng = 1 mg	One million nanograms = one milligram
Units of Volume: The fundamental unit is the *liter*	
1000 L = 1 m³	One thousand liters = 1 cubic meter
1000 ml = 1 L	One thousand milliliters = 1 liter

Question 1: How many micrometers are in a meter?

Question 2: How many centimeters are in a kilometer?

Question 3: How many grams are in a metric tonne (*Tonne* is the correct spelling for the metric unit of 1000 kg)?

Question 4: Express your height in feet, meters, and centimeters.

Question 5: Express your weight in kilograms and pounds.

SCIENTIFIC NOTATION

Very large numbers are most conveniently manipulated (i.e., added, subtracted, multiplied, and divided) by converting the numbers to logarithms to the base 10. Since this is not strictly a math book, we are not going to delve into the theory of logarithms; a practical application is all you need.

The basic fact you need to know is that 10^0 (pronounced "ten to the zero, or ten to the zero power") is defined as 1.

The skills to do the manipulations are easy to learn. The first step is to convert large numbers to scientific notation, with which you should already be familiar. Here's an example. In scientific notation, 18,000,000 is 1.8×10^7.

Here is another example for you to try.

Question 6: Express one billion (1,000,000,000) in scientific notation.

Now let's introduce a wrinkle.

Question 7: Express 2,360,000 in scientific notation.

You can express the same number in a variety of ways using exponents, such as 23.6×10^5, and they all will mean the same thing. But it is customary to express all values in the same format, by placing only one digit to the left of the decimal place (e.g., 2.36×10^6).

Question 8: Express 23,000,000,000,000 (23 trillion) the customary ways using exponents.

For numbers summarized in scientific notation, the prefix in front of the unit (e.g., kilo) denotes the magnitude of the unit (e.g., *kilo*grams = units of 1000 g). *You should memorize these equivalents* (Table 4).

Table 4 ■ Metric Prefixes and Equivalents

Large Numbers		
One thousand	=1000	$=10^3$ (kilo or k)
One million	=1,000,000	$=10^6$ (mega or M)
One billion	=1,000,000,000	$=10^9$
One trillion	=1,000,000,000,000	$=10^{12}$
One quadrillion	=1,000,000,000,000,000	$=10^{15}$ (commonly used in expressions of energy use)
Small Numbers		
One hundredth	=1/100	$=10^{-2}$ (centi or c)
One thousandth	=1/1000	$=10^{-3}$ (milli or m)
One millionth	=1/1,000,000	$=10^{-6}$ (micro or mc or µ)
One billionth	=1/1,000,000,000	$=10^{-9}$ (nano or n)

Question 9: Convert 1.86 mm to (a) nm, (b) µm (micrometers), (c) cm, (d) m, and (e) km. Express your answers as decimals and in scientific notation.

MANIPULATING NUMBERS EXPRESSED IN SCIENTIFIC NOTATION

To add, subtract, multiply, and divide large numbers using scientific notation, you need to remember a few basic rules and check your work carefully. That is all there is to it. Here are the rules.

Multiplication Using Scientific Notation

To multiply numbers expressed in scientific notation, *multiply the bases and add the exponents.*

For example, to multiply:

$$(3 \times 10^3) \times (4 \times 10^5)$$

multiply the bases:

$$3 \times 4 = 12$$

and add the exponents:

$$3 + 5 = 8$$

The result is 12×10^8 or, using the appropriate convention, 1.2×10^9. (NOTE: This is the same as 120×10^7 and any number of other variations as well.)

Division Using Scientific Notation

To divide numbers expressed in scientific notation, *divide the number in the numerator by the number in the denominator and subtract the exponent of the denominator from the exponent of the numerator.* Recall that:

$$\frac{\text{NUMERATOR}}{\text{DENOMINATOR}}$$

For example, to divide:

$$(5.2 \times 10^4) \div (2.6 \times 10^2)$$

divide the numerator by the denominator

$$5.2 \div 2.6 = 2$$

and subtract the exponent of the denominator from the exponent of the numerator

$$4 - 2 = 2$$

So the answer is 2×10^2

Question 10: Perform the following manipulations:

$$(8.7 \times 10^{-3}) \times (4.2 \times 10^{-9}) = ?$$
$$(5.2 \times 10^{18}) \times (8.7 \times 10^{22}) = ?$$
$$(8.7 \times 10^{-3}) \div (4.2 \times 10^{-9}) = ?$$
$$(5.2 \times 10^{18}) \div (8.7 \times 10^{22}) = ?$$

Addition Using Scientific Notation

To add numbers expressed in scientific notation, *you simply add the numbers after converting both to the same exponent.*
For example, to add 3 billion to 14 million:

$$3{,}000{,}000{,}000 + 14{,}000{,}000$$

or

$$(3 \times 10^9) + (14 \times 10^6)$$

convert both numbers to the same exponent (it doesn't matter which, but it is usually easier to use the smaller).

$$3 \text{ billion} = 3000 \text{ million or } 3000 \times 10^6$$
$$14 \text{ million} = 14 \times 10^6$$

Therefore,

$$(3000 \times 10^6) + (14 \times 10^6) = 3014 \times 10^6$$

An equally correct answer would be 3.014×10^9, which is the customary way to express the answer.
Now work out the rule for subtraction using scientific notation.

A WORD ABOUT CALCULATORS

A handheld calculator can help make analyzing environmental issues easy and fun, or it can be a source of great frustration. It all depends on how carefully you put the information into the calculator and how familiar you are with how to use it.

We recommend buying the simplest calculator you can find that will carry out the tasks you need performed. These are typically sold as "scientific calculators." **We also encourage you to take the time to learn how to use the calculator properly.** The calculator should perform the basic data manipulation functions—such as adding, subtracting, multiplying, dividing, determining squares and square roots—and have a single, rather than multiple, memory that is easy to use. Some additional features you should look for include:

- Parentheses ( ) keys
- A y^x key
- A reciprocal **(1/x)** key
- Ability to do simple statistics, including means ($\bar{x}$) and standard deviations (σ)
- An e^x key
- A **LN** (natural log) key
- An **exp** or **EE** key

You should expect to spend around $20 or less.

CRITICAL THINKING[1]

"It is a humanist principle that if you want to know the truth, go to the sources, not to the commentators." Jacques Barzun.[2]

Overview

The mind has awesome power. John Milton said:

The mind is its own place,
And in itself can make
A Heav'n of Hell,
A Hell of Heav'n.

It behooves us all to use that power effectively, which involves critical thinking. Critical thinking is sometimes called *second order thinking*. *First order thinking* is spontaneous, often emotional, and rarely analytical and reflective. As such, it contains prejudice, bias, truth and error, inspiration, and distortions; in short, good and bad reasoning, all mixed together. Second order thinking is essentially first order thinking "raised to the level of conscious realization," that is, analyzed, assessed, and thereby reconstructed.[3]

---

[1]We obtained the basis for much of the information in this section from handouts, discussions, and workshops at the 12th to 20th International Conferences on Critical Thinking and Educational Reform sponsored by Sonoma State University's Center for Critical Thinking and Moral Critique. In addition, we have used *Critical Thinking* by Richard Paul and Linda Elder (see below), an outstanding primer on the subject.
[2]Barzun, Jacques. 2000. *From Dawn to Decadence: 1500 to the Present* (New York: Harper Collins), p. 54.
[3]Paul, R., & L. Elder. 2000. *Critical Thinking—Tools for Taking Charge of Your Learning and Your Life* (Upper Saddle River, NJ: Prentice Hall).

Many scientists equate critical thinking with the application of the scientific method, but we think critical thinking is a far broader and more complex process. Critical thinking involves developing skills that enable you to dissect an issue (*analyze*) and put it together (*synthesize*) so that interrelationships become apparent. It involves searching for *assumptions,* the basic ideas and concepts that guide our thoughts. Critical thinking also encourages an appreciation for our own and for others' *points of view,* which is important when approaching complex environmental issues.

Too often, analyzing complex issues leads some to a belief that everyone is "entitled" to an opinion that should be respected. We do not necessarily concur. However, problem solving demands a willingness to listen for *content* in what others are saying. *Talking is easy, but listening is not.* Developing critical thinking skills is not like learning to ride a bicycle. All of us must learn to use a set of intellectual standards as an "inner voice" by which we constantly test and hone our reasoning skills. But the standards must be set in an appropriate framework in order for true critical assessment to take place.

The following section describes in detail the intellectual standards you should apply when assessing the quality of your reasoning. This is the basis for critical thinking, which in turn is the approach that we try to apply throughout this book.

Intellectual Standards: The Criteria of Solid Reasoning

Clarity This is the "gateway (most important) standard" of critical thinking. If a statement is not clear, its accuracy or relevance cannot be assessed. For example, compare and contrast the following two questions:

1. What can we do about our pollution problem?
2. What can citizens, regulators, and policy makers do to make sure that toxic emissions from industry, transportation, and power generation do not cause irreversible ecological damage, or harm human health?

Accuracy Is that statement true? How can we find out? A statement can be *clear* but not *accurate.*

Precision Can we have more details? Can you be more specific? A statement can be clear and accurate, but not precise. For example, we could say that "There are more sport utility vehicles (SUVs) in the United States in 2001 than ever before." That statement is clear, it is accurate, but how many more SUVs are there? 1? 1000? 1,000,000? (Note that there is a difference between the way many scientists use the word *precision* and the more general way it is used here.)

Relevance How is the statement or evidence related to the issue we are discussing? A statement can be clear, accurate, and precise, but not relevant. Let's take a rather complicated example. If we are given the responsibility to eliminate the harmful environmental impact of pollutants emitted from coal-burning power plants, and we invite public comment on our proposals, someone might say, "Electricity from coal-burning power plants provides power for 100,000 jobs in this state alone." That statement may be clear, accurate, and precise, but it is not relevant to the specific issue of removing pollution (although in other contexts it may in fact be very relevant).

Here is another example: Parks in Arlington, Virginia have signs at all entrances that read: "Dogs must be on leash and under control at all times" and include the relevant County Code citation. Nevertheless, dog owners typically ignore the signs. Here are some of their reasons: "The neighbors don't complain," "There is no dog exercise area near my home," "My dog is well-behaved and doesn't need a leash," or "The police don't mind if our dogs play here." Using critical thinking, assess the relevance of the dog owners' responses.

Breadth   Is there another point of view or line of evidence that could provide us with some insight here? Is there another way to look at this question? For example, you will assess the issue of turfgrass proliferation in Issue 12. In an article one of us (RLM) wrote for the *Washington Post,* he suggested that since turf-care devices such as mowers, trimmers, blowers, and the like are significant sources of air pollution, and since their use is proliferating, it might be easier to address the problem they pose not by banning or regulating these devices, but by reducing the area of turf that must be maintained.

Depth   How does a proposed solution address the real complexities of an issue? Is the solution realistic or superficial? This question is one of the most difficult to tackle, because here is where reasoning, "instinct," and moral values may interact. The points of view of all who take part in the debate must be carefully considered. For example, politicians have offered the statement "Just don't do it" as a solution to the problem of teenage drug use, including smoking. Is that a realistic solution to the problem, or is it a superficial approach? How would you defend your answer? Is your defense grounded in critical thinking?

Logic   How does one's conclusion follow from the evidence? Does the conclusion really make sense? Why or why not? When a series of statements or thoughts are mutually reinforcing and make sense together, and when they exhibit the intellectual standards described above, we say they are logical. When the combination does not make sense, is internally contradictory, or not mutually reinforcing, it is not "logical." Logic in an argument is to the trained mind a bit like the apocryphal definition of obscenity, "You know it when you see it."

Applying Intellectual Standards in a Critical Thinking Framework

The intellectual standards described above are essential to critical evaluation of environmental issues, but there are more factors to be considered. The following criteria constitute the framework in which these standards should be applied.

- **Point of view.** What viewpoint does each contributor bring to the debate? Is it likely that someone who has a job in a weapons plant would have the same view on military spending as someone who doesn't? Would a tobacco company executive be likely to have the same opinion on restricting smoking as someone who lost a relative to lung cancer? Think of other examples, but note that *identifying a point of view does not mean that the opinion should automatically be accepted or discounted.* We should strive to identify our own point of view and the bases for this, we should seek other viewpoints and evaluate their relevance, and we should strive to be fair-minded in our assessment. Few people are won over by having their opinions ridiculed. It is important to note here that our points of view are often informed by our *assumptions,* which we will address below.
- **Evidence.** All problem solving is, or should be, based on evidence and information (sometimes called data, but we prefer to apply this term solely for numbers used in calculations). Our conclusions or claims must be based on sufficient relevant evidence, the information must be laid out clearly, the evidence against our position must be evaluated, and we must be open to new evidence that challenges our conclusions.
- **Purpose.** All thinking to solve problems has a purpose. It is important to have a clear understanding of that purpose and to ensure that all participants are on the "same wavelength." Since it is easy to wander off the subject, it is advisable to periodically check to make sure the discussion is still on target. For example, students working on a research or term project or employees tackling a work-related problem occasionally stray into subjects that are irrelevant and unrelated, although they may be interesting or even seductive. It is vitally important, therefore, that the issue being addressed must be defined and understood as precisely as possible.

■ **Assumptions.** Here is an excerpt from a January 2001 report prepared by the U.S. Energy Information Administration: "With a growing economy, U.S. energy demand is projected to increase 32 percent from 1999 to 2020, reaching 127 quadrillion BTU, *assuming* no changes in Federal laws and regulations." This statement is clear, precise, and contains assumptions.

All reasoning and problem solving depends on *assumptions,* which are *statements accepted as true without proof.* When we assume something, we presuppose it or take it for granted. For example, students show up in class because they assume their professor/teacher will be there. "Never assume" is an old cliche. However, it is more reasonable to be aware of and take care in our assumptions and always be ready to examine and evaluate them. They often need to be revised in the light of new evidence.

Now, before we analyze our own assumptions, let's summarize some characteristics of sound reasoning.

Skilled reasoners:

- Understand key concepts (such as externalities) and ideas.
- Can explain key words and phrases (such as global warming).
- Can explain specific word uses (such as bioaccumulation).
- Continually exercise their thinking skills.
- Can eliminate irrelevant topics and can explain why they are irrelevant.
- Come to well-reasoned conclusions and solutions.

Additionally, the effective reasoner continually assesses and reassesses the quality of his or her thinking in light of new evidence. Finally, the effective reasoner must be able to communicate effectively with others.

Assumptions About Environmental Issues

We cannot stress too heavily the importance and power of assumptions in guiding our reasoning. To give you an example, in the following passages we would like you to respond to the following real-world issues:

First, identify your assumptions in defining government's role in protecting the environment.

Second, determine your assumptions as to the proper *level* of government that may act.

Third, determine your position on the *"precautionary principle."*

Fourth, determine your assumptions concerning the extent to which individuals or institutions in society may impose costs upon others in that society with or without their knowledge or consent.

THE ROLE OF GOVERNMENT

Respond to the following quote, taken from Thomas Jefferson's First Inaugural Address (www.yale.edu/awweb/avalon/presiden/inaug/jefinau1.htm), delivered on March 4, 1801. Most politicians and most Americans probably consider themselves to have "Jeffersonian" principles; for example, former Virginia Governor and now U.S. Senator George Allen quoted this same passage in his own gubernatorial inaugural address to define the principles under which he proposed to govern. Here is what Jefferson said:

> What more is necessary to make us a happy and prosperous people? Still one thing more, fellow citizens—a wise and frugal Government, <u>which shall restrain men from injuring one another</u>, shall leave them otherwise free to regulate their own pursuits of industry and improvements, and shall not take from the mouth of labor the bread it has earned.

Question 11: In a clear sentence or two, explain what you think Jefferson meant by the phrase we underlined.

Question 12: Do you think he was referring solely to thugs who physically brutalize their fellow citizens? Explain.

Question 13: Could he logically have been referring also to citizens who sought to poison others? In other words, is restraining poisoners a legitimate role of government? Explain your answer.

Question 14: Now, what if a citizen or organization dumps a toxin into water or air that all citizens depend on, or if a citizen or organization fills in a wetland that performed valuable ecological functions upon which local residents depend? May government under Jefferson's principle restrain that person or organization?

Your answer to these questions will define your assumptions as to the proper role of government.

THE PROPER LEVEL OF GOVERNMENT THAT MAY ACT

Next, we will ask you to evaluate your assumptions about the *level* of government, if any, that may properly intervene in environmental issues.

One of the major discoveries of the past two decades has been the extent to which much pollution is *transboundary* in nature. For example, one-third of the NOx air pollution affecting the Washington, DC metropolitan area comes from outside the area, from as far away as the Midwest and southern Canada. And, much of the pollution that degrades air quality over Grand Canyon National Park comes from southern California, several hundred kilometers to the west. This situation is repeated over and over across the country.

Question 15:   Therefore, is it appropriate that local government *primarily or solely* bear the responsibility for protecting its own environment? May the states and federal government have a legitimate role based on the transboundary nature of pollution? Explain and justify your answer.

Your answer will help evaluate your assumptions about the extent to which state and federal government agencies have responsibilities to intervene to protect local environments.

THE PRECAUTIONARY PRINCIPLE

Scientists generally define the precautionary principle as follows:

> Action should be taken to prevent damage to the environment even in cases where there is no <u>absolute</u> proof of a causal link between emissions or activity and detrimental environmental effect. Embedded in this is the notion that there should be a reversal of the "burden of proof" whereby the onus is now on the operator to prove that his action will not cause harm rather than on the environment to prove that harm (is occurring or) will occur.[4]

Another way to express this principle is "better safe than sorry." The products of science and technology are often, perhaps usually, brought to the marketplace without adequate investigation into any possible long-term effects on human health and the global environment. Some examples are the uses of lead, mercury, and organochlorines, which you will investigate later.

In most industrialized nations the so-called "burden of proof" falls not on the producers of goods but rather on those who allege that they have suffered harm. This is the basis of our tort system of civil law. As a result of the proliferation of new products, government agencies like the Food and Drug Administration, the EPA, and the Federal Trade Commission, to name but a few, are sometimes unable to keep pace. For example, as of 2001, California had 909 pesticide active ingredients currently registered, used in approximately 11,564 pesticide products. One of six boards and departments within Cal/EPA,

---

[4]Glegg, G. and P. Johnston. 1994. The Policy Implications of Effluent Complexity. In *Proceedings of the Second International Conference on Environmental Pollution* (London: European Centre for Pollution Research), Vol. 1, p. 126. Also see: *Rachel's Environment and Health Weekly*. 1998. The Precautionary Principle. #586, February 19, 1998 (http://www.rachel.org/bulletin/index.cfm?St=2).

the Department of Pesticide Regulation regulates the sale and use of pesticides to protect human health and the environment.

Although individuals have recourse to law if they believe they have been injured, wildlife and ecosystems have no such means of redress. Adherence to the precautionary principle could in the view of many facilitate democratic oversight.

Similarly, a potentially serious threat like global warming, or the proliferation and buildup of organochlorines under the precautionary principle would trigger action to address the threat, even if the "science" is not yet conclusive but is supported by the preponderance of available evidence.[5]

Question 16: Do you accept or reject the precautionary principle? In other words, should those who wish to introduce a new chemical, a new industrial process, a land-use change, and so on, have to demonstrate that their change will not harm the environment *before* proceeding? Explain and defend your answer.

The Question of Externalities

Economists define *externalities* as any *cost* of production not included in the *price* of the good. An example would be environmental pollution or health costs resulting from burning diesel fuel not included in the price of the fuel.[6] Another example would be cleanup costs paid by governments resulting from animal waste degradation of water bodies from large-scale meat processing operations. In this example, the price of chicken or pork at your local supermarket is lower than it would be if all environmental cleanup costs were included in the price of the meat.[7]

Question 17: Consult an economics textbook or do a search on the internet using the term *externalities*. State whether you conclude that externalities should be included in the costs of goods, or whether and in what circumstances some costs can be left for others to pay. Justify your answer using those principles of critical thinking outlined previously.

---

[5]Shabekoff, P. 2000. *Earth Rising: American Environmentalism in the 21st Century.* (Washington, DC: Island Press).

[6]For background on diesel, go to the website of the State of California's Air Resources Board (http://www.arb.ca.gov/homepage.htm).

[7]You can find information on this at the website of Environmental Defense (www.edf.org).

Assumptions About Corporations

Here is another quote from Thomas Jefferson, on the impact of those new organizations called corporations. Corporate power was a focus of the recent candidacy of Ralph Nader for the U.S. presidency. Read Jefferson's words and then respond to the following question.

> I hope we shall take warning from, and example of, England, and crush in its birth the aristocracy of our moneyed corporations, which dare already to challenge our Government to trial, and bid defiance to the laws of our country.

Question 18: Do you share or reject Jefferson's opinions concerning corporations? Cite evidence or provide support to your conclusion. It might help you to prepare a list of positive and negative contributions corporations make to our economy. Do you feel corporations have too much power in contemporary life? Why or why not?

If you are interested in corporate power, you may want to research the 1886 U.S. Supreme Court ruling: *Santa Clara County v. Southern Pacific.* The Court ruled that Southern Pacific was a "natural person" entitled to the protections of the U.S. Constitution's Bill of Rights and fourteenth Amendment.

After researching the case answer this question: Do you believe the Court acted correctly in deciding that a corporation was a person? Is the 1886 ruling relevant to the twenty-first century? Why or why not?

After having thoughtfully responding to the above scenarios, you should now have a better awareness of the assumptions that you bring to the analysis of environmental issues that you are about to undertake.

Summary

To summarize, the intellectual standards by which critical thinking is carried out are clarity, accuracy, precision, relevance, breadth, depth, and logic. These standards are applied in a framework delineated by points of view, assumptions, evidence or information, and purpose. We encourage you to return to this section whenever you need to refresh and polish your critical thinking skills.

USING MATH IN ENVIRONMENTAL ISSUES

The following is an introduction to some of the formulas used in this book.

I. HOW TO PROJECT POPULATION GROWTH USING THE COMPOUND GROWTH EQUATION

For this you will need a calculator with an exponent key.
    The equation is:

$$\textbf{future value} = \textbf{present value} \times (\textbf{e})^{\textbf{kt}},$$

where **e** equals the constant 2.71828 . . . , **k** equals the rate of increase (expressed as a decimal, e.g., 5% would be 0.05), and **t** is the number of years over which the growth is to be measured.

Replacing words with symbols, this equation becomes:

$$N = N_0 \times (e)^{kt}$$

The variable N_0 represents the value of the quantity at time zero, that is, the starting point.

This equation is central to understanding exponential growth and is one of the few worth memorizing. Using it is not as intimidating as you may think.

Sample Growth Calculation Here's an example. Let's figure out (demographers use the word *project*) the world population in 2020, given the mid-year 2000 population of 6.07 billion (6.07×10^9 or 6,070,000,000) and the latest growth rate of 1.25% per year. You can obtain current monthly world population figures at the Census Bureau's World POPClock website (http://www.census.gov/cgi-bin/ipc/popclockw).

Here's how to do this calculation:

$$\textbf{2020 population} = (6.07 \times 10^9) \times e^{(0.0125 \times 20)}$$

How to do this calculation on a typical calculator. On a typical, nongraphics calculator, keystrokes are (commas are for punctuation only):

Key in **0.0125** (the decimal equivalent of 1.25%), then **×** (multiply sign), then **20,** then **=**. This gives you the exponent (the number to which e must be raised). Next hit the button labeled e^x. Note that on most calculators, e^x is labeled *above* a button having another label (frequently ln). If this is the case on your calculator, you must hit the key labeled **2nd** or **2nd F** first, followed by the key with e^x above it. Further, some calculators require you to hit the **=** key after the e^x key.

Next, key in **x** (multiply sign), followed by **the <u>present value</u>** (6.07×10^9; On most calculators this is done by keying in **6.07,** then hitting the button labeled **EE** or **EXP,** then keying in **9.** If the **EE** or **EXP** label is above the key, recall that you must first hit the key labeled **2nd** or **2nd F** before punching **EE** or **EXP.**) Finally, hit the **=** sign.
The correct answer is **7.79 billion.**

Now, project the population of Dhaka in 2056 at a growth rate of 6% per year using the compound growth equation. The 1998 population was 8 million.

$$\begin{aligned} \textbf{2056 population} &= (8.0 \times 10^6) \times e^{(0.06 \times 58)} \\ &= 2.6 \times 10^8 \\ &= 260 \times 10^6 \end{aligned}$$

The compound growth equation can also be rearranged. If you know the starting and ending population sizes over a given period, you can calculate the average growth rate over that period using the formula

$$k = (1/t)\ln(N/N_0).$$

Also, you can calculate how long it would take a population of a given size to grow (or decrease) to a different size at a specified growth rate using

$$t = (1/k)\ln(N/N_0).$$

II. Doubling Time

When a population grows exponentially (by a percentage of the original number), the time it takes for the population to double, called *doubling time* (symbol "**t**"), can be approximately calculated using the following formula:

$$t = (70/k)$$

where t = doubling time (usually in years) and k = the growth rate expressed as the fractional increase or decrease (for example, you would enter 0.07 for a 7% increase).

Derivation of Doubling Time To derive the doubling time formula, we revisit the compound growth equation:

$$\textbf{future value} = \textbf{present value} \times (\textbf{e})^{\textbf{kt}},$$

where e equals the constant $2.71828\ldots$, k equals the rate of increase (expressed as a decimal, i.e., 5% would be 0.05), and t is the number of years (or hours, days, etc.—whatever units you are using in k) over which the growth is to be measured.

Replacing words with symbols, this equation becomes:

$$N = N_0 \times (e)^{kt}$$

The variable N_0 represents the value of the quantity at time zero, that is, the starting point.

If a population doubles in size (that is, increases by a factor of 2), the ratio N/N_0 would be exactly 2. The equation is thus rearranged as follows:

$$N/N_0 = e^{kt}$$
$$2 = e^{kt}$$

Taking the natural log of each side of the equation:

$$\ln 2 = \ln (e^{kt})$$

we get:

$$\ln 2 = kt$$
$$\ln 2 = 0.693, \text{ so:}$$
$$0.693 = kt$$

Dividing by k:

$$0.693/k = t$$

For convenience sake, we round 0.693 to 0.70 and we also multiply the left side of the equation by 100/100, which allows you to enter the rate as a percentage (i.e., 5% would now be entered as 5 instead of 0.05). Thus the final doubling time formula:

$$t = 70/k$$

HUMAN POPULATION GROWTH
HOW LONG WOULD IT TAKE TO FILL UP THE EARTH?

KEY QUESTIONS

- What is the Earth's population and how fast is it growing?
- Where is population growing fastest?
- What is population projected to be in the future?
- How does human population growth threaten the environment?
- What can we do to lessen these threats?
- Who is responsible for solving the problem?

BACKGROUND

Population growth is one of the most contentious subjects in the entire field of environmental studies; it is also basic to any environmental issue.

Why is human population growth an environmental issue? Although humans exert a substantial positive impact with our enormous ingenuity, we also have a profound physical impact on our local, regional, and global environment. First, we take up space—space that at one time was forest, wetland, prairie, or hillside. The space an individual occupies may be minimal, as in cities in New Guinea (Figure 1-1), or it may be large, as in newer U.S. single family houses (Figure 1-2), which averaged over 2100 ft^2 (0.05 acres, or 0.02 hectares) by the late 1990s.[1]

In addition, since humans must eat, we require land for agriculture, and much of that land is irrigated. In most instances this land is also fertilized, usually with industrially produced fertilizers. An additional pollution load is generated by animal waste during meat production in factory farms (also known as Confined or Concentrated Animal Feeding Operations, or CAFOs). The incidence of large-scale fish kills in streams draining agricultural areas and CAFOs is increasing.[2]

People also require transportation. Roads generate polluted runoff, cover permeable land with impervious paving, and take formerly productive land off the tax base. Roads also take an enormous toll on animals (including humans) and other wildlife, divide habitats (which may accelerate species loss), and provide human access to wilderness and for-

---

[1]U.S. Bureau of the Census. 1999. Characteristics of New Housing
(http://www.huduser.org/publications/destech/newhsg99.html).

[2]Spills and Kills: Manure Pollution and America's Livestock Feedlots: A Report by the Clean Water Network, the Izaak Walton League of America, and the Natural Resources Defense Council. 2000.
(http://www.cwn.org/docs/reports/spillkill/spillkillmain.html).

FIGURE 1-1 Human population growth encroaching on the coast at Port Moresby, New Guinea. (© Chris Rainier/CORBIS)

FIGURE 1-2 Typical U.S. beach home in the 2000s. (Chris Major/Moments of Light Photography)

est land, frequently resulting in forest fires and invasion of exotic plant species (by seeds affixing to tires, for example[3]).

The waste created by the growing number of humans is another major environmental problem. In developed countries, a growing amount of human-generated sewage and commercial, residential, and industrial wastes are produced. In the United States, for example, each person produces around 730 kg (1606 lb) of municipal waste per year, not counting sewage, mining, and other industrial waste.[4]

Freer (less-regulated) world trade and a more integrated world economy mean citizens of one country can have an increasing impact on the environment in another. A country can demand raw materials such as tropical hardwoods or ores, and it can ship toxins, in the form of solid waste and air and water pollution, across political boundaries. Nonindigenous species (also called exotic or alien species) have successfully established themselves in estuaries such as San Francisco Bay and Chesapeake Bay, having successfully hitchhiked across oceans in the ballast water of cargo ships and tankers. In their new location, these introduced species can harm ecosystems by disrupting food webs and can cause extensive economic damage—for example, by impacting fisheries or physically clogging water intakes.

So, although each additional human being contributes something unique to the planet and may be the source of problem-solving ingenuity, each one of us places physical demands as well: the more humans, the greater the potential demand. Furthermore, people in developed countries like the United States consume disproportionately more resources than residents of developing countries. Many scientists now conclude that the Earth's ecosystem, the sum of all the planet's smaller ecosystems, may soon reach a point where it can no longer process this demand.

In this issue we are going to examine world population growth patterns, including the impact of AIDS.

EXPONENTIAL GROWTH AND ITS IMPACTS

On October 12, 1999, the United Nations "celebrated" its "Day of 6 Billion," that is, the day at which the earth reached an estimated human population of 6 billion. According to the U.S. Census Bureau,[5] the earth's population was 5 billion in 1987, 4 billion in 1974, 3 billion in 1959, and 1 billion in about 1825 (Figure 1-3).

Before we can analyze the effect of the impact of population growth, it is necessary to understand the simple math that describes such growth.

When any quantity (such as the rate of oil consumption, the human population, or a bank account) grows at a fixed rate (or percentage) per year, say 1% or 10% (as contrasted with growing by a fixed quantity every year, say 500,000 barrels, or 80 million people, or $1000 every year), that growth is said to be *exponential* (Figure 1-3).

Calculating Exponential Growth

Exponential growth, including population growth, is calculated using the compound interest formula (also known as the compound growth equation), the same one used to calculate interest on funds on bank accounts:

---

[3]Macfarlane, G. 1997. Roads and Weeds, Partners in Crime. *The Road RIPorter,* MAY/JUNE 1997: 6–7.
[4]U.S. Census Bureau. 2000. *Statistical Abstract of the United States. Table No. 408* (http://www.census.gov/prod/99pubs/99statab/sec06.pdf).
[5]U.S. Census Bureau. Total Midyear Population for the World: 1950–2050. (http://www.census.gov/ipc/www/worldpop.html).

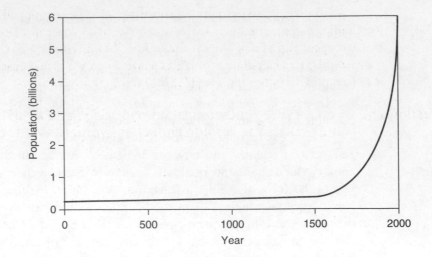

FIGURE 1-3 World population growth.

$$\textbf{future value} = \textbf{present value} \times \textbf{e}^{\textbf{kt}}$$

where **e** equals the constant 2.71828 . . . , **k** equals the rate of increase, and **t** is the number of years (or other units) over which the growth is to be measured.

Replacing the words with symbols, the equation reads:

$$\textbf{N} = \textbf{N}_0 \times \textbf{e}^{\textbf{kt}}$$

The variable N_0 equals the quantity at time zero, i.e., the starting point.

The compound growth equation allows demographers to project future population size, assuming the current population and the population growth rate. Using it is not as intimidating as it may seem at first and is demonstrated in *Using Math in Environmental Issues,* pages 14–15.

In using the compound growth equation to project population growth, we are making two assumptions. First, we assume that our data are accurate. Second, we assume that the population growth, rate (equal to birth rate minus death rate) we are using remains constant over the period we are projecting. Both assumptions may not always be accurate. Population estimates are compiled from censuses all over the world. As you might imagine, making accurate estimates in both developing countries and large nations like China and India is difficult. Growth rates also vary as government policies on family planning and immigration change, along with numerous other variables such as famine, war, and the impact of AIDS. Still, the compound growth equation is a particularly useful tool for examining growth scenarios and planning for the future. Let's practice using it by projecting world population growth.

Question 1: The mid-year 2001 world population was 6.16 billion and was growing at a rate of 1.24%. Project what the world population would be in 2025, 2050, and 2100 at this constant growth rate.

The compound growth equation can also be rearranged. If you know the starting and ending population sizes over a given period, you can calculate the average growth rate over that period using the formula $k = (1/t) \ln(N/N_0)$. Also, you can calculate how long it would take a population of a given size to grow (or decrease) to a different size at a specified growth rate using $t = (1/k) \ln(N/N_0)$.

Question 2: Given a 1987 world population of 5 billion and a 1999 world population of 6 billion, calculate the average annual growth rate over the 12 year period.

Question 3: Given an annual growth rate of 1.5%, how long would it take a population of 5 billion to grow to 10 billion?

Another useful tool for projecting population growth is the *doubling time* formula. For any population that is growing exponentially, the time it takes for the population to double is calculated using $t = 70/k$. (In the doubling time formula, in contrast to the other growth formulas, k is entered as the decimal growth rate $\times$ 100; e.g., a growth rate of 0.07 is entered as 7). The doubling time formula is explained and demonstrated in *Using Math in Environmental Issues,* pages 15–16.

Question 4: Use the doubling time formula to estimate how long would it take a population of 5 billion to double in size given an annual growth rate of 1.5%.

Question 5:   According to the U.S. Census Bureau, world population will increase from 6.16 billion in 2001 to 9.30 billion by 2050. Calculate the average annual growth rate for that period.

Question 6:   Use the growth rate you just calculated in Question 5 to project when the 2001 population will double.

Question 7:   In 1950, 9 countries had a population of 50 million or more. By 1998 this number had grown to 23. Calculate the average annual growth rate of countries of 50 million or more over this 48-year period, then use this rate to project how many countries will have populations greater than 50 million in 2050.

Question 8:   Does a population growth rate of 1.24% sound large? Explain then why it might be cause for concern. Be as specific as you can and cite examples if possible.

Question 9: Biologist Garrett Hardin's "Third Law of Human Ecology" states that the total impact of a population on its environment is determined by the absolute population size multiplied by the impact per person. In our calculations we have thus far ignored the latter. In what ways does your impact on the planet as a U.S. resident differ from your perception of the impact of a resident of a developing country?

Question 10: In light of your answer to Question 9, what do you think would be most effective in minimizing the impacts of humans on the planet: controlling population growth in developing countries (where most growth is in fact occurring) or curbing consumption of resources in developed countries (where the impact per person is highest)? Cite evidence for your answer. Identify the assumptions you used to answer this question.

Question 11: Upholding and improving women's rights and interests (including educational, reproductive, economic, and healthcare) have been called the most important actions to control population growth.[6] Explain how this might work. Do you agree or disagree with the statement? Give evidence for your answer.

---

[6]World Resources Institute. 1995. *World Resources 1994–95* (New York: Oxford University Press).

THE IMPACT OF AIDS

According to the United Nations Population Division.[7] AIDS is enacting a devastating toll on countries in sub-Saharan Africa. They report that life expectancy in the 29 hardest-hit African countries is seven years less than it would have been in the absence of AIDS. In Botswana (Figure 1-4), the hardest-hit country, one of every four adults is infected with HIV, the virus that causes AIDS. Table 1-1 contains both actual (historical) and projected population growth rates for Botswana.

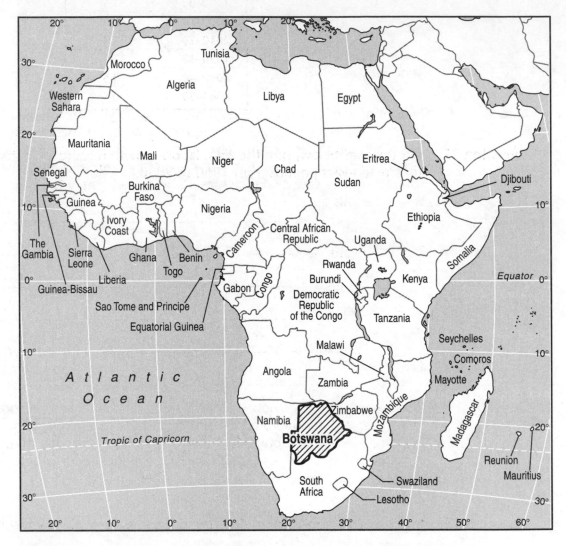

FIGURE 1-4 Map of Africa showing location of Botswana.

[7]United Nations Population Division. The Demographic Impact of HIV/AIDS (http://www.popin.org/pop1998/6.htm).

TABLE 1-1 ■ Average Annual Growth Rates for Botswana for 10-year periods from 1950 to 2050.[8]

Period	Growth Rate
1950–60	1.4
1960–70	1.6
1970–80	4.4
1980–90	3.7
1990–00	1.9
2000–10	−0.5
2010–20	−1.3
2020–30	−1.1
2030–40	−0.5
2040–50	0.5

Question 12: On the axes below, plot the population growth rates (both actual and projected) for the ten-year periods from 1950 to 2050.

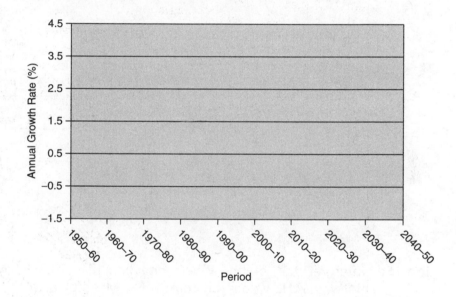

Question 13: Interpret the graph. When does the population growth rate first begin to decline?

[8]U.S. Census Bureau. IDB Summary Demographic Data for Botswana (http://www.census.gov/cgi-bin/ipc/idbsum?cty=BC).

Table 1-2 (below) contains population data for Botswana for the same period.

Table 1-2 ■ Midyear Population Estimates (in thousands) for Botswana for 1950 to 2050[9]

Year	Population	Year	Population
1950	430	1996	1,498
1960	497	1997	1,523
1970	584	1998	1,545
1980	903	1999	1,563
1990	1,304	2000	1,576
1991	1,338	2010	1,502
1992	1,372	2020	1,318
1993	1,406	2030	1,175
1994	1,438	2040	1,114
1995	1,469	2050	1,167

Question 14:   Plot the population data from Table 1-2 on the axes below.

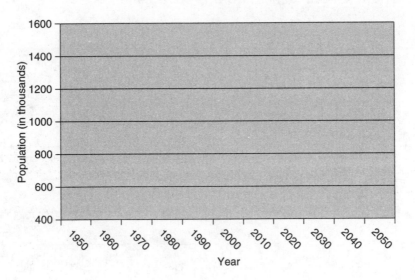

Question 15:   Interpret the graph. When does population actually begin to decrease in Botswana? How does this compare to when the growth rate first began to decline?

Next, we will project population growth in Botswana in the absence of AIDS.

[9]Ibid.

Question 16: Using the highest growth rate in Table 1-1 and beginning with the 1970 population of 584,000, calculate the population of Botswana in 2000 and project it for 2050. Use the formula $N = N_0 \times e^{kt}$. How many more Botswanans would there have been in 2000 in the absence of AIDS? In 2050?

Question 17: Identify and discuss ways in which AIDS might impact Botswana both socially and environmentally.

Question 18: The impact of AIDS on sub-Saharan Africa has been described as devastating. What actions do you think the United States and other developed countries should undertake to lessen the impact of AIDS?

POPULATION DENSITY

Using the modification of the compound interest formula introduced above, $t = (1/k) \ln(N/N_0)$, we could determine when, at a growth rate of 1.24% per year, the world's 2001 population of 6.16 billion would grow to occupy the Earth at a density of one person per square meter (1 person/m^2) of dry land. The earth has 1.31×10^{14} m^2 of dry land. Thus, this number also represents the population size that would fill the earth if one person occupied an area of one meter.

Question 19: Starting with the 2001 world population and using a population growth rate of 1.24%, in what year would the Earth reach this impossible density of 1 person/m^2 of dry land?

Question 20: There are places where population densities approaching 1 person/m^2 already exist. A two-story building in Delhi, India was found to house 518 people[10], a density of 1 person/1.5 m^2. Calculate the floor area of this building in Delhi. How does this compare to the average floor area of a typical U.S. single family home (2100 ft^2 for a new home built in the late 1990s[11])?

Question 21: Identify the problems associated with accommodating large populations by putting people in high-rise buildings and identify any assumptions you used. What point of view did you bring to the question?

---

[10]World Resources Institute. 1996. *World Resources 1996–97—the Urban Environment.* (New York: Author).
[11]U.S. Bureau of the Census. 1999. *Characteristics of New Housing*
(http://www.huduser.org/publications/destech/newhsg99.html).

Question 22: High population densities in urban areas have advantages and disadvantages. List as many of each as you can and defend your conclusion. How could the disadvantages be eliminated or reduced? Use evidence to support your conclusion.

Question 23: Many experts feel that calculations like these transform the discussion about controlling world population from one of *whether* growth will be controlled to one of how and when. Do you agree? Explain why or why not.

For Further Study

- What are some of the implications of human population growth? Consider open space and species loss and how they are affected by the increase in human population. Do you feel that these are causes for concern? Can we solve this problem (if it is a problem) by putting animals in zoos or setting aside land for parks or other green spaces? Why or why not?
- Developing countries are experiencing the most rapid population growth. Is this a problem? Do developed countries have a responsibility to help them control this growth? Specifically, what measures, if any, should be taken?
- How do you respond to those whose opinion differs from yours as to whether population growth is a problem? Make a list of significant points in your argument.
- The developed world's population is aging. Identify the implications. Consider local, regional, national, and global ramifications.
- Herman Daly, a University of Maryland ecological economist, has asserted that population control doesn't deny anyone the "right" to be born, it just asks that they "wait their turn." Do you agree? Discuss and explain your reasoning.

COASTAL GROWTH CASE STUDY: BANGLADESH

KEY QUESTIONS

- Does human population growth threaten coastal areas?
- How can we measure these threats?
- What can we do about them?
- Who is responsible for solving the problem?

Chances are, you live within 100 km (62 mi) of the coast. According to the World Resources Institute,[1] at least 60% of the planet's human population lives that close to the coastline. Coastal areas have the fastest growing populations as well. Not surprisingly, over half the world's coastlines are at significant risk from development-related activities. Some of these activities are:

- Conversion of tropical mangrove communities to fish and shrimp farms (see Issue 22).
- Expansion of many coastal cities that are in the direct path of hurricanes, monsoons, and tropical storms.
- Pollution from Western-style industrial agriculture, including industrial meat production.
- Releases of untreated or partially treated human sewage (between one-third and two-thirds of the human waste generated in developing countries is not even collected).

Coastal cities are reaching sizes unprecedented in human history. Here are but a few examples: Sao Paolo, Brazil, 17 million; Bombay, India, 16.5 million; Lagos, Nigeria, 11 million (and growing at a rate of 5.7% per year); Dhaka, Bangladesh, 8 million (growth rate nearly 6% per year).

---

[1] Bryant, Dirk, et al. 1995. *Coastlines at Risk: An Index of Potential Development-Related Threats to Coastal Ecosystems.* World Resources Institute (WRI) Indicator Brief (Washington, DC: WRI).

BACKGROUND

Bangladesh (Figure 2-1), a country about the size of Illinois, Wisconsin, or Florida, lies on the northern shore of the Indian Ocean. Dr. Selina Begum of the United Kingdom's Bradford University wrote[2]:

> Bangladesh is a delta of the Ganges, Brahmaputra and Meghna Rivers. Tributaries and distributaries [branches] of the river system cover all of the country. The rivers rise in the Himalayas and drain a catchment area of about 1.5 million km^2 [577,000 mi^2], *only 7.5 percent of which lies in Bangladesh* [our emphasis. Stop and think whether the responsibility for Bangladesh's destiny may lie with others far away.].
>
> The country is prone to meteorological and geologic natural disasters, due to its geographic location, climate, variable topography, dynamic river system and exposure to the sea. The steady increase in population continuously increases the potential for natural disaster.

Bangladesh thus makes an ideal case to illustrate the impacts of coastal growth.

Bangladesh was originally part of Pakistan after the Indian subcontinent gained independence from Britain in 1947. In 1970, Bangladesh seceded from the rest of Pakistan. It had a 1999 population of 127 million on a land area of 144,000 km^2 (55,000 mi^2). Bangladesh is vulnerable to catastrophic flooding from river discharges as well as from tropical storms. In addition, it is defenseless against sea level rise from global climate change since more than half the country lies at an elevation of less than 8 meters (26 ft) above sea level (Figure 2-2). As a result, about 25 to 30% of the country is flooded each year, which can increase to 60 to 80% during major floods. Moreover, 17 million people live on land that is less than 1 m (3.3 ft) above sea level.

Flooding in Bangladesh can result from some combination of the following: (1) excess rainfall and snowmelt in the catchment basin, especially in the foothills of the Himalayas (an area where many of the Earth's rainfall records were set); (2) simultaneous peak flooding in all three main rivers; and (3) high tides in the Bay of Bengal, which can dam runoff from rivers and cause the water to "pond up" and overtop its banks.

In addition, changes in land use in the far reaches of the catchment area can result in profound changes in the flood potential of Bangladesh's rivers. For example, deforestation of the Himalayan foothills for agriculture, fuelwood, or human habitation can make the hills more prone to flash flooding. Then, during the torrential monsoon rains, mudslides can dump huge volumes of sediment into the rivers. This sediment can literally pile up in the river channels, reducing channel depth and thereby the channel's capacity to transport water. Result: more water sloshing over the rivers' banks during floods, and more disastrous floods in Bangladesh. Another result: The silt can smother offshore reefs when it finally reaches the ocean. Geologists estimate that the drainage system dumps at least 635 million tonnes (1.4×10^{12} lb) of sediment a year into the Indian Ocean, which is four times the amount of the present Mississippi River.

Some researchers feel that Western-style development is contributing to Bangladesh's increased vulnerability to natural disasters. Farhad Mazhar, director of the Bangladeshi research and development group Ubinin, concludes that the basic problem *is* Western-style development. He cites one example: During predevelopment days Bangladeshi villagers depended on boats for transportation during the monsoon season. But, "to be modern" the Bangladeshis built roads, which blocked floodwaters. These and other "modern improvements" increased flooding from cyclones and have also increased the damage and loss of life. Mazhar concludes, ". . . if you compare the history of the old cyclones and

---

[2]Begum, S. 1996. Climate change and sea-level rise: Its implications in the coastal zone of Bangladesh. *In Global Change, Local Challenge: HDP* (Human Dimensions of Global Environmental Change Programme) Third Scientific Symposium, 20–22 September 1995. Geneva.

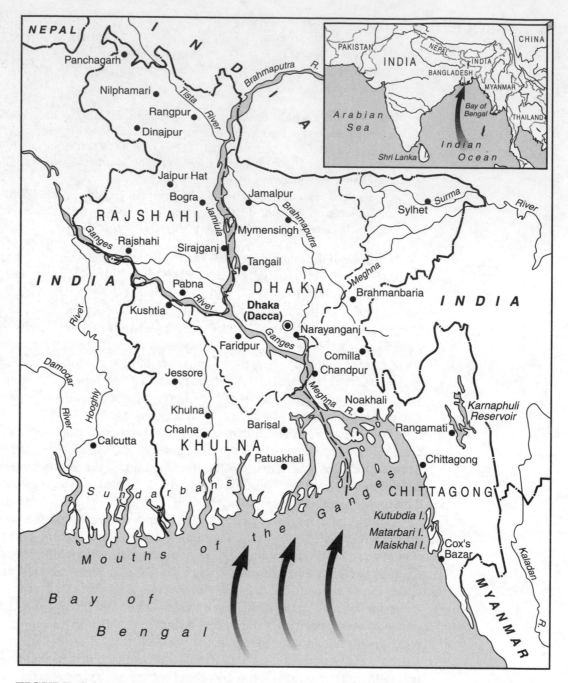

FIGURE 2-1 Location of Bangladesh, showing generalized tracks of major historical typhoons.

flood waters, you'll see . . . the misery has increased with this kind of 'engineering' solution to water."[3]

Truly catastrophic floods can result when high tides coincide with tropical storms. In the early 1990s, such a combination took over 125,000 lives. About one tenth of all Earth's tropical cyclones occur in the Bay of Bengal (Figure 2-1), and about 40% of the global deaths from storm surges occur in Bangladesh.

---

[3]Mazhar, F. What we want from Kyoto. OTN explores global warming. Available: http://www.megastories.com/warming/bangla/kyoto.htm. 6 April 1998.

FIGURE 2-2 High altitude photo of the Bangladesh coastal plain. The entire area is commonly under water during monsoon season. (Photo Disc, Inc.)

Storm surges[4] result when cyclones with winds that can exceed 250 km/hour (155 mi/hr), move onshore, and push a massive wall of seawater onshore with them. Surges in excess of 5 meters (16.4 ft) above high tide are not uncommon, so you can imagine the impact such a storm can have on a country in which half the land is within 8 meters (26.2 ft) of sea level!

Computer models forecast an average rise of sea level of 66 cm (26 in) by 2100 due to global warming, but sea level could rise anywhere from 20 to 140 cm (8–55 in). There is, however, a complicating factor: The deltas on which Bangladesh is built are subsiding, meaning the area will be more susceptible to the flooding that will accompany sea level rise.

Subsidence is a natural feature of deltas. The water in a large river may reach the sea in one or more branches, called *distributaries*. As the distributaries evolve, they may meander more and more, thus lengthening the river's route to the sea. During floods, the distributary can overtop its banks and cut a new and shorter route to the sea. This can result in a rapid shift in where sedimentation occurs at the delta's mouth. An existing delta can sometimes be quickly abandoned, and without the constant addition of new sediment, it subsides.

Here's Dr. Begum[5] again:

A one-meter (3.3 ft) rise in sea level could turn a moderate storm into a catastrophic one. The tropical cyclones of the Bay of Bengal form only when sea surface temperatures reach 27° C or higher. Tropical cyclones in the Bay of Bengal generally form in the pre-monsoon (April–June), and post-monsoon (October–December) seasons. An increase in the surface air temperature (due to global warming) might lead to a more widespread occurrence of cyclones in the Bay of Bengal.

---

[4]For details on storm surges go to the USGS home page at www.usgs.gov.
[5]Mazhar, F. op cit.

Clearly, Bangladesh faces the potential for catastrophe from the combination of coastal population growth, coastal subsidence, land-use changes in the catchment area, and global climate change. As you know, global climate change is thought to be strongly influenced by the release of vast quantities of greenhouse gases, mainly from the burning of fossil fuels in industrial countries and China. Bangladesh's 1992 emissions of CO_2, the principal greenhouse gas, totaled 17.2 *million* metric tons (37.8×10^9 lb), compared to the United States' 4.9 *billion* metric tons (10.8×10^{12} lb), which was by far the world's major source, more than double that of the former Soviet Union. The per capita emissions were even more disparate: 0.15 metric tons (330 lb) per Bangladeshi compared to 19.13 metric tons (42,086 lb) for each resident of the United States.

POPULATION CALCULATIONS: GROWTH IN BANGLADESH

To begin our analysis, we will *project* the population of Dhaka, Bangladesh in the future using the doubling time formula, $t = 70/k$ (where t = the doubling time and k = the growth rate, expressed as %; see *Using Math in Environmental Issues,* pages 15–16 for an explanation and example of how to use the doubling time equation.)

Question 1: The population of Dhaka was 8 million in 1998. First, calculate the doubling time (that is, when the population will increase to 16 million) at the 1998 growth rate of 6% per year.

Question 2: Using the doubling time you just calculated, project what the population will be in five doubling periods (the year 2056).

You can also project population growth using the compound growth formula.

Although a city's or country's population size is a meaningful number, it becomes more meaningful when you know the area (the surface area, that is) in which that population lives. This is known as *population density,* and it is calculated by dividing the population by the area.

Question 3: Bangladesh had a 1999 population of 127 million on a land area of 144,000 km^2 (55,000 mi^2). Calculate the nation's population density in persons per km^2 and per m^2.

Question 4: How does this number compare to the population density of your home state? State population densities are available at the U.S. Census Bureau's website. Go to this website and get data for your home state: (http://www.census.gov:80/population/censusdata/90den_stco.txt).

Question 5: Another way of assessing the potential impact of a country's population on its coast is to calculate the number of persons per meter of coastline (PPMC). The coastline of Bangladesh is approximately 580 × 103 m long[6]. Calculate the number of persons per *meter* of coastline. Use the 1999 population (127 million).

[6]World Resources Institute Facts and Figures—Country Environmental Data (http://www.wri.org/facts/country-data/html).

Question 6: Go to the World Resources Institute's web site (http://www.wri.org/facts/country-data.html), and find the population, total land area, and length of coastline for India, Indonesia, Greece, Chile, Nigeria, and China. Then calculate the population density and the PPMC for each country. Fill in the table below with your data and calculations.

Country	Population	Total Land Area (km^2)	Length of Coastline (km)	Population Density (persons per m^2)	PPMC
India					
Indonesia					
Greece					
Chile					
Nigeria					
China					

Question 7: Which of these—population density or PPMC—do you think is a more meaningful indicator in terms of population impact on the coastal ocean? Why? Use information from your table.

Question 8: What are some of the implications of human population growth for the marine environment? Consider air and water quality, impervious surface, land use, and species loss. Explain why these are or are not causes for concern.

Question 9: We have already established that Bangladesh is a trivial producer of CO_2 and that it faces potential disaster from sea level rise due to global warming and tropical storms. The United States produces nearly a quarter of greenhouse emissions. What responsibility, if any, do you believe the United States has to mitigate (lessen) the potential for environmental disaster in Bangladesh?

Question 10: Assume that the United States is at least partially responsible for the potential for environmental disaster in Bangladesh. List as many actions as you can, by priority (if you can), that the United States could take to prevent, postpone, lessen, or clean up environmental damage and resultant destruction of property and loss of life.

FOR FURTHER STUDY

- Access http://www.population.com/single_category.asp?casefield=4 and compare the current population growth rate for Bangladesh to other coastal countries, including India, Chile, Greece, Indonesia, the Netherlands, Ireland, and Nigeria. These are countries in different climatic zones at different latitudes and also in different stages of development.

POPULATION GROWTH, CARRYING CAPACITY, AND ECOLOGICAL FOOTPRINTS

KEY QUESTIONS

- What factors keep natural populations in balance?
- What does *carrying capacity* mean?
- Can scientists determine the carrying capacity of Earth? Of the United States? Of your home town?

BACKGROUND

In developing his theory of evolution by natural selection, Charles Darwin observed that in any species more individuals are born than will survive to reproduce. If resources are unlimited and environmental conditions are ideal, the number of offspring reaches a maximum. This state of highest reproductive power, which varies widely among different organisms due to genetic and life history differences, is known as a population's *biotic potential*.

However, a number of factors, collectively known as *environmental resistance*, prevents populations from reaching their full biotic potential and thus growing explosively. Environmental resistance includes disease, predation, lack of food, drought, temperature extremes, and other adverse physical or chemical conditions. Most of these factors are density-dependent, that is, their effects are most pronounced when the population density (the number of individuals in a given area) increases beyond a certain level. Thus, the interplay of biotic potential and density-dependent environmental resistance keeps a population in balance.

CARRYING CAPACITY

An important ecological concept related to this balancing act is *carrying capacity*. For a given region, or even globally, carrying capacity is defined as the maximum number of individuals of a given species that the area's resources can support in the long term without significantly depleting or degrading those resources. For humans, this definition is expanded to include (1) not degrading our cultural and social environments and (2) not harming our physical environment in ways that would adversely affect future generations.[1]

---

[1]Bouvier, L., & L. Grant. 1994. *How Many Americans?* (San Francisco: Sierra Club Books).

Determining carrying capacities for most organisms throughout the animal and plant kingdoms is, at least in theory, a fairly direct calculation. However, when carrying capacity is used in a human context, the discussion may become contentious and the resultant calculations subject to dispute. Why is this so?

First, some would argue that the term "human carrying capacity" is meaningless. According to Simon and Kahn,[2] "Because of increases in knowledge, the earth's 'carrying capacity' has been increasing . . . to such an extent that the term . . . has by now no useful meaning."

Second, some scientists point out that air and water pollution do not recognize political boundaries. They thus define the environment occupied by humans and the resources used by them as a global, rather than a regional or national, entity. For them, only a single estimation of human carrying capacity—the Earth's—is valid and meaningful.

Third, carrying capacities for nonhumans are calculated by ecologists using numerical data and mathematical models, but the assessment of human carrying capacities involves not only ecologists, but also economists, policy makers, sociologists, theologians, and so forth. These groups often disagree with each other when it comes to defining what constitutes degradation.

Finally, other organisms are typically limited by their food supply in a particular region, but humans can import food from outside of their region (if one does not consider the earth as a single region). Similarly, humans can export waste and pollution by air and water to areas outside their immediate surroundings. (Some communities even import waste by permitting the construction and operation of huge landfills, i.e., regulated waste dumps, within their community; see Issue 13.) Also, by buying raw materials or manufactured goods from outside their region, some humans can avoid the environmental impact of producing the materials themselves.

POPULATION IMPACT

As you can see, carrying capacity is a complicated subject. Nevertheless, the human carrying capacity of the Earth, or a portion thereof, can be estimated, assuming that different populations have different impacts based on their technology, consumption, and ethics, as well as the simple number of individuals.

For example, 100 million people with a vegetarian diet would have a significantly different environmental impact compared with 100 million people who consume meat. One way to assess this impact is to compare how vegetarian and meat-eating populations affect water supplies. To produce 1 ton of grain, about 1000 tons of water are required. Globally, 40% of all grain goes to meat and poultry production.[3] Thus, high levels of meat consumption put additional stresses on global water supplies.

In terms of energy production and consumption, the impact of industrialized countries varies widely. France, for example, generates over 70% of its electricity by nuclear power and has a significantly different regional energy/environmental impact than China, which generates electricity largely with coal and as of 2001 has few, if any, emission controls on its power plants, though recent evidence suggests this may be changing.

When estimating global or regional human carrying capacity, scientists also study changes that occur in the environment, as well as the rate that these changes occur. Some useful indicators of environmental change include the rate of topsoil loss (see Issue 8), the rate of disappearance of species (see Issue 24), the rate of degradation of water quality, and the rate of change in the composition of the atmosphere (see Issue 18).

---

[2]Simon, J., & H. Kahn. 1984. *The Resourceful Earth.* (Malden, MA: Blackwell Publishers).
[3]Postel, S. 1996. *Dividing the Waters: Food Security, Ecosystem Health, and the New Politics of Scarcity.* (Washington, DC: Worldwatch Institute).

And finally, carrying capacity may often have less to do with population density and ultimately more to do with the impact of a society's technology, resource demand, and waste generated.[4]

CARRYING CAPACITY AND ECOLOGICAL FOOTPRINTS

In mid-2001 the population of the United States was about 281 million.[5] Are our natural resources successfully supporting 281 million people? How do the imported goods that maintain our level of economic activity and standard of living affect the global environment? How fast is our population growing? Will our natural resources be able to support a growing population? If so, for how long? Which segment of our population is growing the fastest? How much of our growth is due to legal immigration? Illegal immigration?

Obviously, the exact human carrying capacity of an area is very difficult to determine, especially for a country as large and diverse as the United States. Despite the difficulty, it is essential that estimates of human carrying capacity be made so that policymakers can act to ensure that our environment is able to support human life and natural biodiversity into the future.

Question 1: Examine the items in the following list. For each one, write a few sentences supporting the argument that it is a carrying capacity issue.

a. Die-offs of squirrels on your campus

b. Implementation of a Fishery Management Plan for the east and Gulf coasts of the United States limiting catches of sharks

---

[4]McConnell, R.L. 1995. The human population carrying capacity of the Chesapeake Bay watershed: A preliminary analysis. *Population and Environment 16:* 335–351.
[5]U.S. Bureau of the Census U.S. POPClock Projection (http://www.census.gov/cgi-bin/popclock).

c. The loss of 70,000 km² of cropland each year due to nutrient depletion

d. Increasingly heavy vehicular traffic and "road rage"

e. Near-record high prices for heating oil and natural gas in recent winters in the northeast United States and elsewhere

f. Rapidly rising real estate prices in California and coastal communities

g. The loss of 50% of wetlands in 22 states since colonial times

Question 2: Now, for each item, explain why it might *not* be a carrying capacity issue—that is, list alternative explanations.

  a. Die-offs of squirrels on your campus

  b. Implementation of a Fishery Management Plan for the east and Gulf coasts of the United States limiting catches of sharks

  c. The loss of 70,000 km^2 of cropland each year due to nutrient depletion

  d. Increasingly heavy vehicular traffic and "road rage"

e. Near-record high prices for heating oil and natural gas in recent winters in the northeast United States and elsewhere

f. Rapidly rising real estate prices in California and coastal communities

g. The loss of 50% of wetlands in 22 states since colonial times

Question 3:   If there had been fewer humans, which of the above items might have changed? How does this change your opinion of whether these are carrying capacity issues?

Question 4: Write down several (or as many as you can) indicators, environmental or otherwise, that may support the assertion that the United States had exceeded its carrying capacity. Be as specific as you can. Example: Logging of old-growth forests to meet demands for hardwoods.

Question 5: It has been stated that everyone in the United States could fit comfortably inside the state of Texas. The mid-2001 United States population is 281 million. The area of Texas is 261,914 sq. mi. (6,780,000 ha). Calculate how many square miles, acres, and hectares each person would occupy if all U.S. residents lived in Texas (1 mi^2 = 640 acres; 1 acre = 0.4047 ha).

SUSTAINABILITY

To live sustainably in an area, some would argue that human populations must remain at or below that area's carrying capacity. *Sustainability* is defined as "meeting the needs of the present without compromising the ability of future generations to meet their own needs,"[6] or, more specifically, not using resources faster than they can be replenished. However, as we have already discussed, humans in one area are able to import food, water, and manufactured products from other areas and export waste to other areas. These are typically not factored into carrying capacity estimates. Another way of examining the impact of humans is to estimate the overall impact each of us has, a measure known as our *ecological footprint*. Every individual has an ecological footprint that extends well beyond the geographical area in which that person lives. *The Ecological Footprint of Nations*[7] presents estimates for how much of the earth's area we appropriate for our needs. The average American, for example, uses 10.3 ha (25.4 acres) to support his or her lifestyle. This includes farmland, forests, mines, dumps, schools, hospitals, roads, playgrounds, malls, etc. (Figure 3-1).

Question 6: How many times larger would Texas need to be, assuming all Americans lived there and each American required 10.3 ha?

[6]United Nations World Commission On Environment and Development (The Brundtland Commission), *Our Common Future*, 1987.

[7]Wackernagel, M. et al. (http://www.ecouncil.ac.cr/rio/focus/report/english/footprint/).

FIGURE 3-1 Aerial photo of a mall in Charleston, SC. To comply with local zoning regulations, shopping centers must have a minimum number of paved parking spaces, which contributes to the large ecological footprint of Americans. (Courtesy of South Carolina Coastal Conservation League)

Question 7: The surface area of the Earth is 51 billion hectares. Assuming that it is all usable to serve as a footprint, calculate how much area is available for each of the 6.16 billion inhabitants in 2001. Report your answer in square miles, hectares, and acres.

Question 8: Now assume all 6.16 billion people lived like Americans. How much area would be needed? Report your answer in square miles, hectares, and acres. How many more planets the size of Earth would be required?

Question 9: How much area would be required for all 6.16 billion people if they lived like the 142 million residents of Pakistan, who each require 0.8 ha? How many more planets the size of Earth would be required?

In fact, not all of the Earth's surface can serve as footprint area. Approximately 6.4 billion hectares are marginally productive or unproductive since they are covered by ice or lack water.[8] Additionally, 36.3 billion hectares are seawater.

Question 10: Summarize what you have learned about carrying capacities and ecological footprints.

Question 11: Americans use a disproportionate portion of the earth's resources. Discuss reasons why this is so.

Question 12: Discuss whether the use of resources by Americans is a fair allocation of the planet's resources.

---

[8]Ibid.

Question 13: Is our use of resources sustainable? Why or why not? Cite specific examples and document your assertions with evidence.

Question 14: Because estimating an area's human carrying capacity is inexact and difficult, some would argue that the concept is useless or not worthy of discussion. Do you agree or disagree? Explain and justify your answer.

Question 15: In their book *How Many Americans?*, Bouvier and Grant[9] estimate the human carrying capacity of the United States at 150 million. List evidence that either supports or refutes their claim and discuss your conclusions.

FOR FURTHER STUDY

- A number of websites will allow you to estimate your ecological footprint. Conduct a search on the web using the key words "Ecological Footpint." Once you have accessed a suitable site, calculate your footprint. How does your ecological footprint compare to others? What actions could you take to reduce your footprint?
- Many architects, planners, developers, builders, and citizens are reducing their ecological footprints by using sustainable community and building design. Research this topic on the web. Are there individuals or groups in your community involved in planning or building sustainably?

---

[9]Bouvier, L., *op. cit.*

Issue 4

POPULATION GROWTH, MIGRATION, AND FERTILITY

KEY QUESTIONS

- What are fertility rates and how are they determined?
- How much does immigration contribute to population growth in the United States?
- Can the environmental impacts of immigration be quantified?

FERTILITY, MIGRATION, AND POPULATION GROWTH

Annual population growth rates[1] in the world range from -1.1% in the Republic of Latvia to $+8.6\%$ in Liberia; the average annual growth rate of the United States from 1995 to 2000 was 0.8% per year. What factors contribute to these growth rates? Are growth rates rising or falling? What impact will these have on world population? Does a growth rate of 0.8% sound high or low? We will address these questions in this issue.

Our current growth rate of 0.8% would result in the U.S. population doubling to over 550 million in about 88 years (see *Using Math in Environmental Issues,* pages 14–16 for a discussion of doubling time). Among industrialized countries, this is a high rate. For example, the average annual population growth rates for the same period were 0.3% in Germany and 0.1% in the United Kingdom.[2] Overall, world population growth rates are declining, although global population is not expected to stabilize until after the year 2200.[3]

For a population to increase, there must be more births than deaths and/or greater immigration than emigration. Let us first examine births, which demographers refer to as *natality* or *fertility*. These may be expressed several ways. The easiest, called the crude birth rate (CBR), is expressed as the number of births per 1000 people and is found by dividing the total births for the year by the mid-year population, then multiplying by 1000. Globally, CBRs[4] range from 9 to 10 in low-natality countries like Germany and Italy to over 50 in high-natality countries like Niger, Jamaica, and Iran. The crude birth rate for the United States, averaged for the period 1995 to 2000 was 13.8 live births per 1000 people, which is somewhat higher than other industrialized countries but fairly low compared with the rest of the world. Population growth can be easily projected using crude birth and death rates.

---

[1] Averaged for the period 1995-2000. World Resources Institute
(http://www.wri.org/wr-98-99/pdf/wr98_hd2.pdf).
[2] Ibid.
[3] United Nations Population Division (http://www.popin.org/6billion/t01.htm).
[4] Averaged for the period 1995-2000. World Resources Institute
(http://www.wri.org/wr-98-99/pdf/wr98_hd2.pdf).

Question 1: Project growth in the U.S. population at a crude birth rate of 13.8 and crude death rate of 8.8 by completing the following table, starting at the 1998 population (270,248,000). For this calculation we will ignore the data for migration. (NOTE: A computer spreadsheet program such as Excel or Lotus would be ideal for performing the manual repetitive actions involved in this calculation.)

To make this calculation, first subtract the crude death rate (8.8) from the crude birth rate (13.8).[5] Then multiply this number by the beginning population divided by 1000. (NOTE: Some calculators will not allow you to enter numbers this large without using scientific notation.) This quantity allows you to calculate the natural rate of population growth, that is, growth that doesn't count migration.

Year	Initial Population	Net Addition to Population (to nearest hundred)	Total Population at End of Year
1998	270,248,000	1,351,000	271,599,000
1999	271,599,000		
2000			
2001			
2002			

Given the beginning and ending population levels, it is simple to calculate a growth rate. To do so, take the difference between the ending and starting values and divide by the starting value. Multiply by 100 to give you the growth rate in percent. Also, recall that you can use the formula $t = (1/k) \ln(N/N_0)$ (see *Using Math in Environmental Issues,* pages 14–15) to find the answer.

Question 2: Calculate the growth rate of the United States using the above table. (Recall that this rate ignores the data for migration.)

---

[5]This quantity is also known as the *natural increase.*

Question 3: A passbook bank account earned around 2% interest in 2001, whereas the interest rate for a home mortgage was around 7.5%. Compared to these, does the growth rate you just calculated sound large or small? Why?

Question 4: If a stable population is our goal (still ignoring the data for migration), then we must decrease the birth rate to equal the death rate. By what percentage must the crude birth rate be reduced to equal the death rate?

Question 5: Suggest policies that might reduce birth rates in the United States. Which are the most feasible? Why? Which are the least feasible? Why?

Although both the crude birth and death rates are simple to calculate and thus easy to use, a better measure of fertility is *total fertility rate* (TFR), the average number of children that a woman will bear over her reproductive life. Again ignoring migration, for the U.S. population to stabilize, the TFR should equal about 2.1.[6] The TFR of new immigrants (>2) is significantly higher than that of residents who are not recent immigrants (<2).

MIGRATION AND POPULATION GROWTH

In the preceding questions we calculated that the population of the United States would grow annually at 0.50% at the 1995–2000 crude birth and death rates if we ignored the data for migration. Now let's calculate the rate of growth of the U.S. population using the data for migration. Emigration (out-migration) is relatively small (<200,000 annually); legal immigration is approximately 700,000 annually (in 1998, it was 660,000, the lowest since

---

[6]A TFR of 2.1 allows for a low rate of infant and child mortality. Thus, in countries with high infant and child mortality rates, the replacement TFR is substantially higher.

FIGURE 4-1 Haitians crowd a boat at Port-au-Prince in an attempt to reach the United States (©David Turnley/CORBIS).

1988[7]). The number of immigrants illegally entering the country is difficult to estimate (Figure 4-1). Conservatively, many experts estimate net immigration (legal and illegal immigration minus emigration) at between 1 and 1.3 million.[8] We'll use the smaller number.

Question 6: Use your calculator or computer spreadsheet program to fill in the table below, assuming the previous crude birth and death rates and net immigration of 1 million per year.

Year	Initial Population	Net Addition Births Minus Deaths	Immigration	Population at End of Year
1998	270,248,000	1,351,000	1,000,000	272,599,000
1999	272,599,000		1,000,000	
2000			1,000,000	
2001			1,000,000	
2002			1,000,000	

Question 7: Calculate the annual population growth rate from the preceding table.

---

[7]U.S. Department of Justice. Office of Immigration and Naturalization. 1999. Legal Immigration Fiscal Year 1998 (http://www.ins.usdoj.gov/graphics/publicaffairs/newsrels/98Legal.pdf).
[8]U.S. Department of Justice. Office of Immigration and Naturalization (http://www.ins.usdoj.gov/graphics/aboutins/statistics/illegalalien/index.htm).

Question 8: For 1998, what percentage of total population growth was due to immigration? Discuss whether you think that immigration makes a significant contribution to U.S. population growth.

Now, let's examine the impact of reducing birth rate on population growth under three different immigration scenarios.

Question 9: Assuming that the net addition to population is approximately half of what it was in 1998 (i.e., 2.5 instead of 5.0 per 1000), fill in the three population tables below.

CASE A: Birth rate–death rate = 2.5 per thousand
Immigration = 1,000,000

Year	Initial Population	Net Addition: Births	Net Addition: Immigration	Population at End of Year
1998	270,248,000			
1999				
2000				
2001				
2002				

CASE B: Birth rate–death rate = 2.5 per thousand
Immigration = 500,000

Year	Initial Population	Net Addition: Births	Net Addition: Immigration	Population at End of Year
1998	270,248,000			
1999				
2000				
2001				
2002				

CASE C: Birth rate–death rate = 2.5 per thousand
Immigration = 150,000

Year	Initial Population	Net Addition: Births	Net Addition: Immigration	Population at End of Year
1998	270,248,000			
1999				
2000				
2001				
2002				

Question 10: Plot population at end of year for the above three cases on the same graph (below).

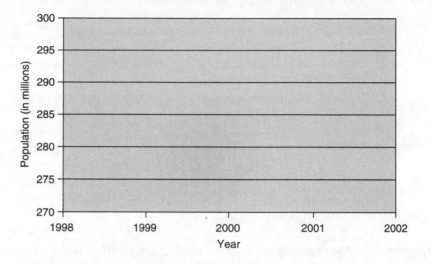

Question 11: Calculate population growth rates for each case. What do you conclude about the roles of fertility and immigration in U.S. population growth?

Question 12: List the positive contributions that you believe immigrants make to the United States. Identify any assumptions you used (i.e., diversity is desirable, etc.).

Question 13: Discuss ways in which the United States might be affected by reducing immigration. List some advantages and disadvantages (e.g., economic, social, political, and cultural gains/losses) and justify your choices.

A report issued in 1997 by the National Research Council[9] at the request of the U.S. Commission on Immigration Reform attempted to quantify the impact of immigration on U.S. society. Among its conclusions were:

- If immigration continues at present levels, it will account for two-thirds of the U.S. population growth to 2050.
- Immigrants are more poorly educated than residents, thus imposing significant costs on localities.
- Immigration lowers wages among the less-skilled residents. It is responsible for nearly 50% of wage depression among low-skilled U.S. workers since 1980.
- Immigration is a contributing factor in the widening gap between the rich and the poor in the United States.
- Immigration imposes considerable cost on U.S. society with little economic benefit. Immigrants may add $1 to $10 billion annually to U.S. gross domestic product (GDP; the total annual value of goods and services), but they cost $15 to $20 billion per year in government services.

Thus, at a time of constricting government responsibilities and flat or reduced outlays for social services, over 1 million immigrants per year apparently impose significant costs on U.S. society, with much of that cost falling on those least able to pay.

Question 14:   From an environmental perspective, immigrants to the United States quickly begin to consume energy and natural resources at the same rate as residents. Respond to the argument that immigration to the United States should be limited to conserve energy and natural resources. Is it reasonable to support immigration reductions for this reason without seeking to reduce Americans' own disproportionate consumption of global resources?

For Further Study

- Research migration patterns within the United States using *Statistical Abstract of the United States*.[10] In terms of environmental implications, how does this differ from immigration from foreign countries?
- Look up your state's population in *Statistical Abstract of the United States*.[11] How many equivalents of your state are added to the U.S. population annually?
- Research the impact of immigration on other countries (such as Germany, France, and Israel) and how these countries are responding.
- Access the summary of the McCarthy and Vernez[12] report on the costs and benefits of immigration on California and discuss the impact of immigration on that state.

---

[9]National Research Council. 1997. *The New Americans: Economic, Demographic, and Fiscal Effects of Immigration* (Washington, DC: Author).
[10]http://www.census.gov/statab/www/.
[11]Ibid.
[12]McCarthy, K.F. and G. Vernez. 1997. Immigration in a changing economy: California's experience. Available from RAND: order from http://www.rand.org/cgi-bin/Abstracts/e-getabbydoc.pl?MR-854.

enues to pay for their occupant's demands in county services.[6] As a result, many counties actively encourage "town house" construction far from urban centers. Research the transportation implications of such decisions.

■ Find U.S. Census data (www.census.gov) on growth for your county and make similar calculations to those we did for Prince William. (The California Department of Finance[7] has excellent statistics, for example.) What is the growth rate? What are its implications? Consider transportation. According to the *Statistical Abstract of the United States* (1999),[8] there are approximately 0.6 motor vehicles per person in the United States Assuming the number of motor vehicles grows with the population, assess the impact of population growth on road congestion in your county or city. Is road construction and maintenance an unavoidable cost of population growth? Should people who choose not to own cars pay for these new or expanded roads? Provide evidence for your conclusions.

■ Examine your local zoning ordinances, which you can usually find at your city or town planner's office. Determine if they are designed to promote or limit growth.

■ Study articles from your local newspaper, or from a metropolitan newspaper in your library, that focus on growth. Are growth rates given? If not, can you calculate them? Do the articles describe growth as "healthy" or "robust"? Do you agree that growth is "healthy" or "robust"? Discuss. Many of these articles refer to high population growth as "population booms" or "baby booms." What is the connotation of these descriptions? Can you think of a term that is neutral or more appropriate?

■ To find out what your state is doing about growth, go to http://www.smart-growth.org/information/news/news_trends12-00.html.

---

[6]Behr, P. 2000. Taking no prisoners in battle over growth. *Washington Post*, Oct. 30, 2000.
[7]http://www.dof.ca.gov/.
[8]http://www.census.gov/statab/www/.

LOCAL POPULATION GROWTH II:
CASE STUDIES, MARYLAND AND COLORADO

KEY QUESTIONS

- What are "smart growth" initiatives designed to accomplish?
- What are brownfields?
- What is Maryland's "Smart Growth" Initiative?
- What is Colorado doing to control growth?

Many states are becoming concerned about the implications of uncontrolled growth. Two approaches that we illustrate here are Maryland's more aggressive management style and the less aggressive style of Colorado.

MARYLAND'S "SMART GROWTH" INITIATIVE

In 1997 the Democratic Governor of Maryland, Parris Glendening, proposed a "smart growth" initiative.[1] Here is how the state described its goals:

- To save the most valuable remaining natural resources before they are forever lost.
- To support existing communities and neighborhoods, using state resources to support development in areas where the infrastructure (roads, sewers, etc.) is already in place or planned to support it.
- To save taxpayers millions of dollars in the unnecessary cost of building the infrastructure required to support sprawl.

One of the first steps Maryland took to fulfill the goals of the initiative was the designation of "Priority Funding Areas," which are locations where the state and local governments chose to encourage and support economic development and new growth. Priority Funding Areas included (1) every municipality, (2) areas inside the Washington Beltway and the Baltimore Beltway, and (3) areas already designated as enterprise zones, neighborhood revitalization areas, heritage areas, as well as existing industrial land.

State policies also sought to:

1. Facilitate the reuse of "brownfields" (a term for abandoned, usually urban, sites that were often contaminated with environmental pollutants). Often, unused or abandoned properties that are contaminated, or even thought to be contaminated

---

[1]http://www.op.state.md.us/smartgrowth/index.html.

by potential commercial or industrial developers, are not attractive for development because of understandable uncertainty about potential future liability. These sites can often be cleaned up to a level suitable for commercial and industrial uses and, importantly, are already served by infrastructure such as water and sewer. However, because of these liability concerns, developers and businesses usually prefer to locate instead on "greenfields"—pristine farms and open spaces. These areas often lack infrastructure such as roads and utilities. This process on the one hand contributes to the loss of farms and open spaces, and on the other hand needlessly increases the amount of taxpayer dollars spent on funding new infrastructure. Furthermore, greenfield siting impedes revitalization efforts in rundown neighborhoods.

2. Provide tax credits to businesses creating jobs in a Priority Funding Area.[2]
3. Create a "Live Near Your Work" pilot program that provided cash contributions to workers buying homes in certain older neighborhoods.[3]
4. Encourage more preservation of undeveloped land by creating a Rural Legacy Program (RLP) to provide financial resources for the protection of farm and forest lands and their conservation from development. Here is some of what the state said about the RLP: "The Program will reallocate State funds to purchase conservation easements for large contiguous tracts of agricultural, forest and natural areas subject to development pressure . . . private land trusts will be encouraged to identify Rural Legacy Areas and to competitively apply for funds to complement existing land conservation efforts or create new ones."[4]

Question 1:   List reasons why governments and citizens should care about preserving farm and forest land. (Hint: Start by thinking about "watersheds[5]". Most of Maryland is within the Chesapeake Bay Watershed.)

Question 2:   Summarize the aspects of the Maryland "Smart Growth" program. Especially address how it treats "property rights" of those landowners living in critical areas.

---

[2]For information go to http://www.op.state.md.us/smartgrowth/taxcredit.htm.
[3]For details, go to http://www.op.state.md.us/smartgrowth/lnyw.htm.
[4]For funding sources, go to http://www.op.state.md.us/smartgrowth/legacy.htm.
[5]A bay's watershed is the area drained by streams that feed the bay.

Question 3: How does this initiative deal with "sprawl"?

COLORADO AND SPRAWL DEVELOPMENT

Colorado is one of the nation's fastest growing states. Locals claim the growth is fueled by "refugees" from overgrown and polluted California and Texas. Whatever the reason, Colorado is experiencing unprecedented threats to its heralded quality of life: Indeed, some say it may be in the throes of being "loved to death."

In 1999, Republican Governor Bill Owens explained the Colorado plan as follows: "We are the stewards of Colorado's future. For the sake of our children and grandchildren, we must preserve Colorado's natural beauty and provide opportunities for future generations to pursue their own dreams. Our task is to protect our special Colorado way of life."[6]

The Colorado strategy has four major components, called "initiatives." They are:

Initiative One. Natural Landscapes, Saving Open Space, Ranches and Farms, which includes

- Conservation Easement Purchases or Leases. Conservation easements keep the land in agricultural production while allowing farmers and ranchers to realize some of the value of their land. Farmers and ranchers may voluntarily sell or lease the development rights to conservation groups and state agencies and receive credits to reduce their state income tax bills.

- Governor's Commission on Saving Open Space, Ranches, and Farms. The Commission examines the effectiveness of efforts to protect open space, and evaluates strategies to enhance the preservation of open space, ranches, and farms. The Commission makes recommendations to the Governor, state legislature, local governments, and private organizations to help them most effectively spend the tens of millions of dollars each year spent to save open space in Colorado.

- Wildlife Habitat Preservation Tax Credits. This program rewards landowners who preserve large parcels of their land in its natural state for use as wildlife habitat. A credit to reduce state income tax bills is provided for landowners that the state certifies as managing their land in ways that provide critical wildlife habitat.

- Wildlife Habitat Landstrips. Planting narrow strips of native grasses and shrubs along fence lines, roads, and streams often provides critical habitat for birds, fishes, and small animals. The Colorado Department of Highways is exploring reduced mowing along road right-of-ways to encourage wildlife habitat landstrips. In addition, a number of incentives are being explored for private landowners that provide wildlife habitat landstrips along their roads, fence lines, and streams. All these methods are voluntary.

---

[6]http://www.state.co.us/gov_dir/govnr_dir/11-29-99a_speech.htm.

- Brownfields Tax Credit. This provides state sales tax and state income tax relief for developers who rehabilitate and renovate dilapidated areas in cities and towns. The tax credit encourages the clean up and redevelopment of unsightly and/or polluted urban "brownfields" rather than encouraging development on "greenfield" sites far from existing infrastructure.
- Acquiring Land for New State Parks. In 2000 construction was scheduled to begin on Great Plains State Park, the first state park in the Southeastern Plains portion of the state. Nearly 1700 acres of land will also be acquired for Cheyenne Mountain State Park, in southern Colorado Springs.

Initiative Two. Strong Neighborhoods: Protecting Our Way of Life

The Office of Smart Growth in the Department of Local Affairs will coordinate the state's efforts to assist local communities as they plan for and manage growth. These efforts include:

- Colorado Heritage Planning Grants with matching local funds will be available for adjacent communities to better plan for growth.
- Colorado Heritage Reports. The Office issues Colorado Heritage Reports on innovative strategies and best practices in the areas of land use planning, intergovernmental agreements and open space preservation.[7]

Initiative Three. Opportunity Colorado: Bringing Prosperity to the Whole State of Colorado

Parts of Colorado have experienced growing incomes over the past decade. This prosperity is primarily concentrated along the heavily populated I-25 corridor from Fort Collins to Colorado Springs and along the I-70 corridor from Denver to Eagle. With the exception of a few pockets elsewhere, such as Aspen and Durango, much of the rest of the state was described as "struggling with low incomes, inadequate job creation and diminished economic opportunities." This initiative has the following components:

- To assist Coloradans with home purchases, the state has a Home Ownership Tax Credit. The "Colorado Dreams" program provides tax relief for businesses that assist lower income workers and their families with home closing costs or down payments. Businesses must provide a match to the employee's contribution in order to receive the credit.
- The state also provides a Colorado Dreams Low Income Housing Tax Credit. The Colorado Dreams program provides tax relief for developers who build or renovate low-income rental housing. Low-income housing is a critical problem in wealthy areas like Aspen, in which housing costs are so high that low-income workers, and some public employees, cannot afford housing in the community. This requires many workers to drive long distances to keep their jobs.

Initiative Four. Moving Forward: Creating Our Transportation Future

This initiative deals with transportation and has two categories, only one of which is directly related to addressing sprawl. Colorado has more than 85,000 miles of roads and 8,300 bridges. In 1999, vehicle miles traveled in Colorado totaled more than 36 billion. The state proposes support for mass transit such as light-rail where population densities

---

[7]For examples, go to http://www.state.co.us/smartgrowth/download.html.

warrant, where projects are "clearly viable and affordable options" to improve air quality and reduce congestion. One example cited is along I-25/I-225 in downtown Denver.

Question 4:   Assess the state's commitment to controlling population growth and sprawl development.

Postscript: In the 2000 election, Coloradans defeated Amendment 24, an initiative sponsored by the Colorado Public Interest Research Group and the Sierra Club, which would have changed the state's Constitution by requiring communities to map existing growth areas currently served by water and sewer infrastructure. These maps then would have been used to manage growth.

Sprawl in Denver

We will next consider a report published by the environmental group Environmental Defense, titled "Sprawl in the Denver Region."[8] In a 1999 survey, 60 percent of Denver residents identified growth, traffic, and sprawl as the region's most serious problems.

Since 1990, the Denver region has grown at a rate of 2.22% per year. It has a present area of 550 square miles (1425 km^2) and a 2000 population of 2.38 million. Since 1950, Denver's population has increased by 200%, while land consumed for development has grown 350%.[9]

Question 5:   Environmental Defense says: "Colorado is fortunate: more than one third of the state's land area is in public ownership, managed by the federal or state government." Do you agree with this position? Explain your position using critical thinking principles (see *Basic Concepts and Tools* for review).

---

[8]http://www.edf.org/pubs/Reports/DenverSprawl/.
[9]http://www.edf.org/pubs/Reports/DenverSprawl/DenverSprawlReport.pdf.

Question 6: Determine the doubling time, t, for Denver's population at its present growth rate. If necessary, refer to *Using Math in Environmental Issues* pages 15–16 to review how to calculate doubling time.

Question 7: Determine the doubling time for the projected population growth rate for 2000–2020 of 1.9%.

Question 8: Calculate when the Denver area's population will reach 4.76 million, using both the current and projected growth rate.

According to Environmental Defense, one way to characterize whether a community is in "balance" is to compare jobs to residents.

A community with a healthy balance of jobs and housing is one where people can both work and afford to live. With this balance comes the possibility of fewer people commuting, and less regional traffic congestion and air pollution. Municipalities that fail to achieve a balance of jobs and housing do so for many reasons. It can happen when a local government allows residential development without adequate commercial development, or when residential development is limited but commercial development is encouraged.

Accounting for children and those not looking for work, a ratio of 0.6 to 0.85 (i.e., 0.6 to 0.85 jobs per resident) is believed to indicate a "balanced" community. Lacking an integrated mass-transit system (which is missing in the Denver area), higher values will bring unwanted auto commuters into the community, and lower values will "drive" residents elsewhere to look for work. So, imbalance will unnecessarily add road costs, which residents are increasingly unwilling to pay for.

FOR FURTHER STUDY

- Go to the Environmental Defense report (http://www.edf.org/pubs/Reports/DenverSprawl/DenverSprawlReport.pdf) and scroll to page 24, which depicts a graph of the ratio of jobs per resident for incorporated areas within the Denver Metropolitan Area in 1996. List those communities which fall within the "balanced" ratio. (Remember, values too low are just as undesirable as values that are too high.)
- Describe growth and development in your home town and in the town in which your college or university is located. Are you aware of any "smart growth" planning or initiatives? Describe growth policies that you would support.

Issue 7

GLOBAL GRAIN PRODUCTION: CAN WE BEEF IT UP?

KEY QUESTIONS

- How did humans domesticate wild grains?
- What are global trends in grain production?
- What are the principal uses of grain in the United States and the world?
- Can food supply keep up with a growing population?

WHY WE EAT

Reduced to its basic chemical components, the human body is simply carbon, hydrogen, oxygen, nitrogen, and a few other elements. These combine to form more complex, highly organized structures called molecules, such as proteins, lipids, carbohydrates, and nucleic acids. These molecules interact to build even larger and more complex structures: cells, tissues, organs, organ systems, and ultimately, the individual.

Such a system of highly organized, integrated parts, however, does not exist without cost. It is subject to the same laws of physics that affect any piece of matter. Among these is the *second law of thermodynamics,* which states that disorder (entropy) in the universe is increasing and that whenever energy is converted from one form to another, some of the input energy is lost in the process of conversion, resulting in less output energy. Practically speaking, in the natural state, complex, organized systems tend to become simple and unorganized. They do, that is, unless energy is added to the system to keep it organized and complex.

What constitutes appropriate nutrition for a human? First, we need energy, which is measured in kilocalories (Cal).[1] A typical person needs between 1500 to 2500 calories (i.e., kilocalories) per day. In addition, we need vitamins, minerals, and proteins. Inadequate nutrition leads to undernourishment or malnutrition. Globally, perhaps 1 billion people are underfed.

WHAT WE EAT

The human diet generally is omnivorous; this means that humans are able to eat from all the major food groups. Wheat, rice, and maize (called corn in the United States) are the three major cereal crops used by humans. Additional food resources include potatoes, bar-

---

[1]A calorie is the amount of heat required to raise the temperature of 1 g of water by 1 degree C. A kilocalorie = 1000 calories. Unfortunately, the nomenclature is confused because the "calories" that we count when we eat are actually kilocalories.

ley, oats, cassava, sweet potatoes, sugar cane and beets, legumes, sorghum, millet, vegetable oils, fruits, and vegetables, along with meat (beef, poultry, pork, mutton), milk, fish, and other seafood.

GRAINS AND THE ORIGIN OF AGRICULTURE

Humans are extremely dependent on a few grains: among them are corn (maize), wheat, rice, and barley. Why do we depend so heavily on these relatively few plant species? For one thing, most wild plants that we could possibly domesticate are useless to us as food—they may be indigestible, very low in food value, or hard to prepare. In fact, most wild biomass is wood or leaves, which we can't digest.

Plant domestication has been defined as "growing a plant, and thereby consciously or unconsciously causing it to change genetically from its wild ancestor in ways that make it more useful to humans."[2] The history of plant domestication is a fascinating one, which we briefly summarize.[3] Here's how domestication may have come about. Studies of plant remains at human habitation sites tell us when humans first domesticated them, since domesticated grains are much different from their wild ancestors. According to Carbon-14 dates, grain production probably arose first in the "Fertile Crescent" of the Near East, with the domestication of wheat from wild varieties occurring by 8500 BCE.[4] It quickly spread along the great east-west axis of Eurasia. The domestication of rice began in China sometime around 7500 BCE. In the "New World," corn (maize) was first domesticated by 3500 BCE.

How did domestication arise? Probably as a result of decisions made by early farmers, who were still also hunter-gatherers, without being aware of the consequences of such decisions. In other words, farmers didn't plan to be farmers.

Grain domestication was preceded by a major change of climate that resulted from the last withdrawal of glaciers perhaps 11,000 years ago. This warming led to an expansion of the habitat for the wild grains that were ultimately to be domesticated, making it easy for humans to collect them. At the same time, technology was advancing, resulting eventually in tools that would allow the more effective tilling of soil and the breaking up of tough sods.

When humans began to grow crops from wild varieties they had collected, they would naturally select the varieties that would best suit their needs. Intuitively, we might conclude that when the first "farmers" began to plant crops, they would have collected those they liked the best—the biggest, tastiest varieties they could find. Recall that there is a wide variation in appearance of wild fruits like blueberries, blackberries, and strawberries. Generally, early farmers probably selected according to size—the bigger the fruit, the bigger the offspring, or so they may have thought. Cultivated peas, for example, are ten times bigger than their wild ancestors, which hunter-gatherers must have been collecting for centuries. To take another example, the oldest corn cob from Mesoamerica was only 1 to 2 cm long. By 1500 farmers were growing 15 cm varieties, and modern cobs may exceed 30 cm.

Domestication of the plants we take for granted did not of course happen overnight. Rather, it was a process that likely developed over centuries. In fact, it is still occurring. Strawberries were not domesticated from wild forms until the Middle Ages, and pecans were not domesticated until the middle of the nineteenth century. And some of the ancestors of the plants we cultivate were poisonous! Examples are eggplant, potatoes, and almonds. These plants do have a *recessive* gene, however, that will not grow poisonous offspring. In the wild, such a gene is likely to stay recessive, since animals will likely gobble

---

[2]Diamond, Jared. 1997. *Guns, Germs and Steel* (New York: W.W. Norton), p. 114.
[3]Ibid.
[4]"BCE" means "before the common era" and means the same as BC.

up any edible varieties. But humans, after a bit of admittedly risky trial-and-error, would eventually select plants with the recessive gene and begin to grow only forms that displayed the nonpoisonous characteristics they needed.

Let's close this brief history by giving you an illustration of the value of recessive genes to human society. Wild wheat plants have a dominant gene that causes the plant's stalk to disintegrate when the plant's seeds are ripe. This allows the seeds to be spilled upon the ground, permitting them to germinate. But a recessive variety exists that lacks this gene: These stalks don't disintegrate, and this is the wheat plant that early farmers selected.

Now, to summarize, here is why we became dependent on so few grains:

- Their ancestors were edible and plentiful in the wild.
- They grew quickly and were easily collected and stored.
- They contained protein and were rich in carbohydrates.

Thus, as a result of millennia of trial and error, by the days of the Roman Empire almost all modern grains were being cultivated in some form somewhere.

THINKING CRITICALLY ABOUT GLOBAL GRAIN PRODUCTION

The total area of land used for agriculture rose from 4.55 billion hectares in 1966 to 4.93 billion in 1996. Over most of human history, farmers have increased agricultural output mainly by plowing up forests and natural grasslands. Limits of geographic expansion were reached decades ago in densely populated parts of India, China, Java, Egypt, and Western Europe. *Intensification* of production (obtaining more output from a given area of agricultural land) has thus become a "growing" necessity. In some regions, particularly in Asia, this has been achieved primarily through producing several crops each year in irrigated "agro-ecosystems" using new, fast-growing crop varieties—the so-called "Green Revolution."

But our efforts to increase agricultural production are becoming more challenging and more costly. There is evidence that growth rates in grain yields have slowed in both developed and developing countries. And future increases in food production may become more difficult because a complex set of environmental and social factors must now be taken into account while developing new crop technologies. An illustration of the latter is the growing controversy over genetically modified foods.

Let's start our analysis by considering the following quotations from Bailey[5]:

Food is more abundant and cheaper today than at any other time before in history. Per capita grain supplies have increased 24 percent since 1950, while food prices have plummeted by 57 percent since 1980.

Food production has outpaced population growth since the 1960s. The increase in food production in poor countries has been more than double the population growth rate in recent years.

Now consider these two quotations from Brown et al.[6]:

Grain output that easily outpaced population growth for more than 30 years now lags well behind.

Headline: GRAIN STOCKS DROP TO ALL-TIME LOW World carryover stocks of grain for 1996 are projected to drop to 229 million tons, down from 296 million tons in 1995.

---

[5]Bailey, R. (ed.) 1995. *True State of the Planet* (New York: Free Press), p. 50.
[6]Brown, L., et al. 1996. *State of the World* (New York: Worldwatch Press), p. 79.

Question 1: List your conclusions, based on these statements.

Question 2: Are the two sets of statements consistent with each other? What additional information would you need in order to choose which statement(s) is/are accurate?

What are the facts about world grain production? Is it keeping pace with population growth? What do projections look like? The following discussion deals with calculations germane to these and other questions.

GRAINLAND AREA

In the United States, many rural and suburban dwellers are used to seeing farms being converted into residential subdivisions with miles and miles of tract homes serviced by highways and power lines. At the same time, there has been intensification of agricultural land use around some major foreign cities (and to an unexpected extent, within cities), particularly for high-value perishables such as dairy and vegetables, but also to meet subsistence needs.

Globally, is the area devoted to grain production actually decreasing? In 1950, 587 million hectares (remember there are 2.47 acres per hectare) of grainland area were harvested. For 1999, around 674 million hectares of grainland area were harvested.[7]

Question 3: What was the percentage increase in grainland area harvested from 1950 to 1999?

---

[7]Worldwatch Institute. 2000. *Vital Signs 2000* (New York: W.W. Norton).

Question 4: Using the formula $k = (1/t)\ln(N/N_0)$ (see *Using Math in Environmental Issues*, pages 14–15 for an example of how to use this formula), calculate the annual percentage increase in grainland area harvested from 1950 to 1999.

Question 5: Refer to the two sets of quotations on p. 75. Which one do your calculations tend to support? Explain your reasoning.

Question 6: By 1981, world grainland area had steadily increased from 1950 and reached a peak at 732 million hectares. From 1982 to the present, the area has shrunk. Calculate the percentage increase in world grainland area from 1950 to 1981.

Question 7: Calculate the annual percentage increase in world grainland area from 1950–1981.

Question 8: The world's population in 1950 was approximately 2.55 billion. In 1999 it was 6.0 billion. Calculate the per capita area of grainland harvested for 1950 and 1999.

Question 9:   How do the last three calculations change your perception of which set of quotations is more accurate? Why?

Although the per capita grainland area harvested has decreased by nearly 50% from 1950 to 1999, this decrease has been offset by compensatory increases in output per hectare achieved by high-yield farming. Termed the "Green Revolution," the development and use of new strains of wheat and rice along with greater irrigation (Figure 7-1), input of fertilizers, pesticides, herbicides, fungicides, and so forth, drastically increased crop yields.

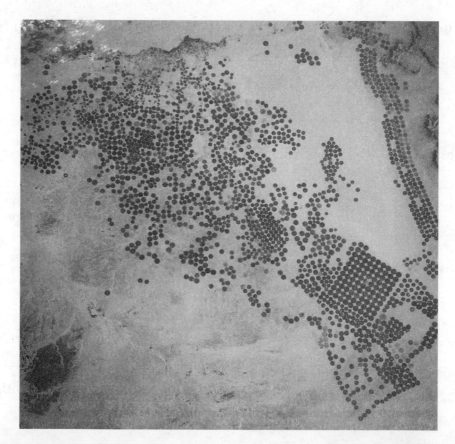

FIGURE 7-1 High altitude space photo of irrigated fields in the Arabian desert. Irrigation water is from 1300 m deep aquifers. Circles are approximately 0.5 miles (550 m) in diameter (CORBIS).

GRAIN AND MEAT PRODUCTION

In 1950, world grain production was 631 million tons. In 1999, it had risen to about 1855 million tons.[8]

Question 10: Calculate the percentage increase in grain production over the period 1950–1999.

Question 11: Calculate the annual percentage increase in grain production over the period 1950–1999.

Question 12: Determine the per capita grain production for 1950 and 1999.

Question 13: How do the last set of calculations change your perception of which set of quotations on p. 75 is more accurate?

---

[8]Ibid.

Question 14: Dennis Avery[9] states that the lags in global grain production in the early 1990s were due to "affluent countries . . . trying to limit their surpluses and because of chaotic conditions in the former Soviet Union." Other explanations include bad weather, declines in global fertilizer use, and a failure of irrigation and high-yield crop varieties. In light of these explanations, do you think that conditions favorable to expanding per capita grain production can be counted on in the future? Explain the bases for your answer.

Question 15: World meat production reached an all-time high in 1999. Economic growth in Southeast Asia has fueled this increase in part. Globally, annual per capita meat consumption ranged from 115 kg in the U.S. to less than 1 kg in India in 1999. Over 15 billion livestock (11 billion of which are poultry) exist at any one time to satisfy this demand[10] (Figure 7-2). Does the increase in meat production necessarily mean a more equitable distribution of meat than previously? In other words, does the increase in meat production imply that global nutrition is improving? Explain why you chose your answer.

FIGURE 7-2 Cattle feedlots near Lubbock, TX. (©Richard Hamilton Smith/CORBIS)

---

[9]Bailey, R. (ed.) 1995. *True State of the Planet* (New York: Free Press).
[10]McKinney, M.L., & R.M. Schoch. 1996. *Environmental Science: Systems and Solutions* (Minneapolis/St. Paul: West Publishing).

Question 16: In addition to grasses not suitable for human consumption, much of the world's grain harvest is fed to cattle and other livestock. In 1960, 294 out of 822 million tons of grain were used as animal feed. In 1995, estimates were that 644 out of 1750 million tons were used as animal feed. In the United States, of the 740 kg of grain (excluding exports) that was produced per person in 1990, 663 kg were fed to livestock.[11] Calculate the percentage of global grain harvest fed to livestock for the years 1960 and 1995. Make the same calculation for the United States for 1990.

Question 17: What are alternative uses for this grain?

Question 18: Pimentel and Pimentel[12] estimate that a switch to more grass-fed livestock in the United States would allow about 130 million tons of grain to be diverted to human consumption, which could feed about 400 million people annually. Using the Pimentels' estimates for the United States, if none of the global grain production were fed to livestock, how many more people could be fed globally?

Question 19: Should we discourage meat consumption, or encourage feeding more grass instead of grain to livestock, to free up grain to feed the world's hungry? Explain your answer.

---

[11]Pimentel, D., & M. Pimentel. 1996. *Food, Energy, and Society* (Niwat, CO: University Press of Colorado).
[12]Ibid.

Question 20:   Prime farmland is disappearing around large U.S. cities. The land is being used for housing, roads, office buildings, and shopping centers. Some have questioned whether it is "economical" to convert farmland around cities into enormous parking lots for grocery stores in which food grown thousands of miles away is sold. Do you concur? Discuss and defend your answer using critical thinking principles.

FOR FURTHER STUDY

■ Research the consequences of our penchant for meat. Discuss the environmental impact. An excellent source is Jeremy Rifkin's book, *Beyond Beef: The Rise and Fall of the Cattle Culture.*[13]

■ Research the effect a shift from grain-fed to grass-fed beef would have on public lands. This is a significant political issue, especially involving the prices charged ranchers to graze livestock on public lands. You can find information on this by conducting a search on the Internet, by consulting the websites of the U.S. Department of the Interior and the U.S. Bureau of Land Management.

■ Demand for organically grown food is increasing in rich countries, at rates that approach and sometimes exceed 20% per year. In agriculture, alternatives to toxic pesticides, herbicides, and fungicides are available. Growing numbers of farmers are going completely pesticide-free, profiting from consumers who as of 2000 spend $22 billion a year on organic products. Research the issue at www.worldwatch.org.

---

[13]Rifkin, J. 1993. *Beyond Beef: The Rise and Fall of the Cattle Culture* (New York: Dutton).

SOIL DEGRADATION AND EROSION

KEY QUESTIONS

- What is the status of the world's soils?
- What can be done to ensure fertile soils in the future?
- What is the relationship between soils and atmospheric CO_2?

Soils, said Leonardo da Vinci, are the Earth's flesh. They are critical to the survival of human society. But we are just beginning to understand the role of soils in maintaining a healthy planet. Soils may have natural regulatory roles and benefits undreamed of by humans. For example, recent discoveries show that soils contain a wide range of natural antibiotics produced by teeming soil microbes. The antibiotic streptomycin was itself discovered by soil investigators.

Some pedologists (soil scientists) are growing alarmed at the extent to which humans are affecting and changing soils, all in a very short time, compared to the time it takes soils to form. (Soils vary greatly in age and rate of formation. While some soils may be millions of years old, most are younger.) Pedologists wonder if they will survive in their natural state long enough for us to truly understand them.

WHAT ARE SOILS?

Soils (Figure 8-1) are composed of decomposed and disintegrated bits of rock and mineral matter (e.g., insoluble residues like iron hydroxides that water can't dissolve), humus, or organic matter, and perhaps most importantly, uncountable soil organisms, mostly microscopic, and largely undescribed by science. In the words of one pedologist, soils are "living, breathing entities,"[1] and while they are very important to agriculture, they are perhaps of greater significance to society at large.

SOIL DEGRADATION

Humans degrade soils by

1. Paving soils over with asphalt and buildings (urbanization)
2. Accelerating soil erosion
3. Poor agricultural practices that result in desertification

---

[1]Schneiderman, J.S. 2000. *The Earth Around Us: Maintaining a Living Planet* (New York: W.H. Freeman), pp. 152, 153.

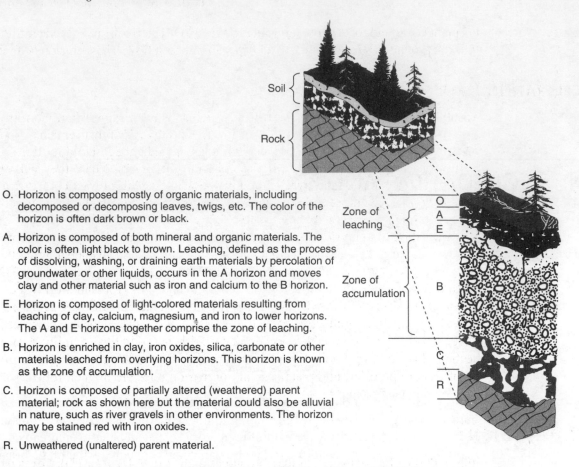

O. Horizon is composed mostly of organic materials, including decomposed or decomposing leaves, twigs, etc. The color of the horizon is often dark brown or black.

A. Horizon is composed of both mineral and organic materials. The color is often light black to brown. Leaching, defined as the process of dissolving, washing, or draining earth materials by percolation of groundwater or other liquids, occurs in the A horizon and moves clay and other material such as iron and calcium to the B horizon.

E. Horizon is composed of light-colored materials resulting from leaching of clay, calcium, magnesium, and iron to lower horizons. The A and E horizons together comprise the zone of leaching.

B. Horizon is enriched in clay, iron oxides, silica, carbonate or other materials leached from overlying horizons. This horizon is known as the zone of accumulation.

C. Horizon is composed of partially altered (weathered) parent material; rock as shown here but the material could also be alluvial in nature, such as river gravels in other environments. The horizon may be stained red with iron oxides.

R. Unweathered (unaltered) parent material.

FIGURE 8-1 The structure and composition of soils. (Edward A. Keller)

4. Irrigation practices leading to salt buildup
5. Possibly by adding toxins to kill soil microbes

URBANIZATION AND SOIL LOSS

Urbanization may ultimately be a greater threat to soil loss than agriculture. In testimony before Congress, American Farmland Trust (AFT) President Ralph Grossi said, "For the last 60 years America has sought to prevent the erosion of topsoil from wind and water, and through concerted action we have for the most part succeeded. Yet we continue to lose soil—to asphalt. In fact, the volume of topsoil lost to urbanization is roughly equivalent to the soil saved by the (federal government's) Conservation Reserve Program!"

And the AFT estimates that, based on land-use changes in the Central Valley of California—America's most important agricultural region—by 2050 one-third of that region will be paved over with suburbs.

Professor Ronald Amundson of UC Berkeley wrote, "we are appalled when a 3000-year-old tree is cut down, but our culture generally applauds the ingenuity of developers who can transform a 300,000 year-old soil into a golf course or housing tract."[2]

In terms of the impact of topsoil loss on estuaries like Chesapeake Bay, sprawl development produces from five to seven times the sediment of a forest and nearly twice as much sediment as compact development. Forests allow little topsoil to erode, by contrast with conventional agriculture. In undeveloped forested land, (1) runoff is minimal,

---

[2]Ibid.

(2) plant cover and roots allow rainwater to seep slowly into soil, and (3) water filters slowly through ground to stream, which results in stable stream flow. Thus, erosion is minimal.

ACCELERATING RATES OF SOIL EROSION

Troubling many pedologists is the *rate* at which soils are disappearing. Pioneers reported soils 16 feet (5 m) deep in the prairies of Iowa, but only half of that remains. Floods in the midwestern states over the past several decades have dumped untold millions of tons of fertile soils from farms into the Mississippi and other rivers, where they will eventually end up behind dams, on the river's flood plain, or in the Gulf of Mexico. Hurricane Floyd in 1999 similarly dumped millions of tons of topsoil into the Atlantic from North Carolina farms and hog processing operations (see Figure 5-4, p. 60).

According to a report from the World Resources Institute,[3] the greatest source of top-soil loss along with sprawl development is from improperly managed agricultural land. By 1990, poor agricultural practices had contributed to the degradation of 562 million hectares, about 38% of the roughly 1.5 billion hectares in cropland worldwide. Some of this land was only slightly degraded, but an appreciable amount was damaged severely enough to impair its productive capacity or to take it out of production completely. Since 1990, losses have continued to accumulate, with an additional 5 million to 6 million hectares lost to severe soil degradation annually. Soil erosion often results from water runoff due to poor farming practices, and if erosion rates are not controlled, soil erosion can degrade water supplies.

DESERTIFICATION

Introduction of certain agricultural practices may cause long-term or irreversible land degradation. The introduction of the steel plow to the U.S. Great Plains broke up the tough soil, removed the native long-rooted prairie grasses, and introduced cattle. All of these processes led to the onset of an episode of wind erosion unprecedented in the region's recent history: the so-called "dust bowl" years of the 1930s (Figure 8-2). Similar practices have led to the degradation of vast tracts of the African "Sahel," an area immediately south of the Sahara.

SALT BUILDUP

In many arid regions, soluble elements such as selenium and sodium may accumulate in soils since there is insufficient rainfall to wash them deeper into the water table. Selenium, which is a nutrient in minute quantities, is toxic at higher concentrations. If such land is brought into agricultural production by large-scale irrigation, the selenium may be mobilized and washed into rivers, where it may poison wildlife. In the Central Valley of California for example, groundwater conditions are such that irrigation waters may pond up near the land surface, eventually poisoning the soils with toxins like selenium. Indeed, the site of the original domestication of grains, the Fertile Crescent in what is now Iraq, may have been permanently poisoned by intense irrigation practices.

ADDING TOXINS

Cultivated soils are different from uncultivated soils. Strawberry and wine grape growers like to "sterilize" their soils to rid the soils of soil pests like nematodes, which can feed on the roots of young plants and thus reduce yields. One such chemical in widespread use,

---

[3]http://www.igc.org/wri/trends/soilloss.html.

FIGURE 8-2 Loss of topsoil during Dust Bowl era, Guyman, OK, 1937. (©Bettmann/ CORBIS).

methyl bromide, has been banned in many countries because of threats to the health of agricultural workers, and because the chemical is one of the most potent ozone-destroying chemicals known to the planet's upper atmosphere ozone-shield.

SOILS AND AGRICULTURAL PRODUCTION

One of the greatest threats to agricultural production, according to experts, is the loss of topsoil (Figure 8-2). Most farmers would agree with Kansas State University's extension service, that "the challenge is to manage soils so they will be as productive for future generations as they are today."

Most topsoil loss—about two-thirds—is caused by rains washing away topsoil, with another third caused by wind. One analysis of global soil erosion estimates that, depending on the region, topsoil is currently being lost 16 to 300 times faster than it can be replaced. Soil-making processes are notoriously slow and may require from 200 to 1000 years to form 2.5 centimeters (one inch) of topsoil.

But farmland can be degraded in other ways besides erosion. Mechanical tilling can lead to soil compaction and crusting. Repeated cropping without sufficient fallow (rest) periods or replacement of nutrients with cover crops, manure, or fertilizer can deplete soil nutrients. The loss of topsoil directly affects agricultural productivity (Figure 8-3).

But research shows that "no-till" agriculture can result in drastically reduced soil erosion compared to other forms of tillage[4] (Figure 8-4).

Oregon State University's Extension Service reports that no-till agriculture coupled with abandoning the traditional "summer fallow" system can substantially limit topsoil erosion. The following is from a report prepared by OSU's Extension Service.[5]

There's a quiet revolution going on in the wheat fields of northeastern Oregon. Growers of dryland wheat are experimenting with planting crops every year, an approach that can save thousands of tons of topsoil.

For nearly 100 years, most growers used a summer fallow system. That is, land planted with wheat one year was plowed but left unplanted the following year to give the soil time to

---

[4]Kansas State University, Great Plains Dryland Conservation Technologies. Cooperative Extension Service, Manhattan KS (http://www.oznet.ksu.edu/library/crpsl2/S81.pdf).
[5]http://osu.orst.edu/dept/ncs/newsarch/1999/Apr99/crop.htm.

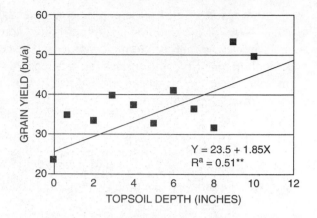

FIGURE 8-3 Impact of topsoil loss on grain yield. (Havlin, J. L., H. Kok, & W. Wehmueller. 1992. Soil erosion—Productivity Relationships for Dryland Winter Wheat, pp. 60-65 in J.L. Havlin (ed.), *Proc. Great Plains Soil Fertility Conf.*, Denver, CO, Mar. 3–4, 1992. As cited in: Enhancing Agricultural Profitability and Sustainability. Kansas State University Extension Service. (http://www.oznet.ksu.edu/library/crpsl2/S81.pdf).

build up enough moisture to produce another crop. But this practice results in massive erosion, such that organic matter in the soil is now one-half its 19th century level.

In the early 1990s, the Columbia Basin had 625,000 acres in summer fallow a year. Annual cropping was virtually zero. By 2000, 16% of the traditional summer fallow acreage—more than 100,000 acres—was being planted every year.

Annual cropping has become feasible thanks to changing practices such as better timing and application of fertilizer, new pesticides, and improved methods for planting crops without plowing.

Beginning in 1995, OSU Extension conducted a number of demonstration projects with annual crops such as spring barley, canola, and lentils. Other demonstrations involved no-till farming, the practice of planting seeds directly in the stubble of the previous crop, eliminating the need to plow the soil.

Planting a crop every year dramatically reduces soil erosion. On high yielding wheat fields, it cuts soil erosion from 12 tons per acre under the fallow system to 6 tons.

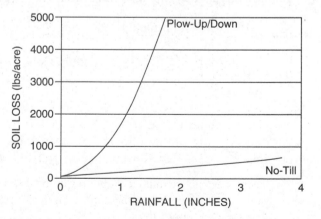

FIGURE 8-4 Soil loss reduced by no-till agriculture, Columbia basin, Oregon. (Dickey, E., P. Harlan, & D. Vokal. 1981. Crop residue management for water erosion control. NebGuide G-81-554. University of Nebraska-Lincoln, NE. As cited in Enhancing Agricultural Profitability and Sustainability. Kansas State University Extension Service (http://www.oznet.ksu.edu/library/crpsl2/S81.pdf).

As a result of OSU research, non-summer-fallow crop acreage could reach 250,000 acres—40% of the traditional summer fallow acreage—by 2007, although much will depend on weather and economic conditions. The economic benefits of annual cropping are difficult to quantify, since prices farmers receive vary considerably depending on the crop (mainly wheat). With wheat prices down in world markets, growing crops every year offers growers a product other than wheat to sell in more favorable domestic and "niche" export markets.

Even more important, however, is the future of farming in the Columbia Basin. The erosion and soil deterioration associated with summer fallow practices cannot continue for much longer. Annual cropping provides an alternative to a wasteful farming practice that will, if continued, eventually put farmers out of business in the Columbia Basin.

TRENDS IN SOIL DEGRADATION

According to scientists at the International Food Policy Research Institute (IFPRI), by mid-2000 nearly 40% of the world's agricultural land was seriously degraded, which could undermine the long-term productive capacity of those soils.

The evidence compiled by IFPRI suggests that soil degradation has already had significant impacts on the productivity of about 16% of the globe's agricultural land. Their study concludes that almost 75% of crop land in Central America is seriously degraded, 20% in Africa (mostly pasture), and 11% in Asia.

Question 1:   In geopolitical terms, how could such land degradation affect U.S. population growth? (Hint: Consider immigration and refugees.)

Threats to the world's food production capacity are compounded by three disturbing trends, according to experts at the IFPRI:

1. 1.5 billion additional people will be on the planet by 2020, almost all in poorer developing countries.
2. The natural fertility of agricultural soils is generally declining.
3. It is increasingly difficult to find productive new land to expand the agricultural base.

Question 2: Suggest solutions to the three issues raised above and cite evidence for your choices.

One of the most common management techniques used to maintain the condition of agro-ecosystems is the application of inorganic fertilizers (nitrogen, phosphorus and potassium) or manure. Too little can lead to soil "nutrient mining" (amount of nutrients extracted by harvested crops is greater than the amount of nutrients applied), and too much can lead to nutrient leaching (washing away of excess nutrients contaminating ground and surface water). Nutrient depletion can result in severe water pollution.

IFPRI experts have overlaid maps of nutrient balance for the countries of Latin America and the Caribbean with trends in yields to identify potential degradation "hot spots," where yield growth is slowing and soil fertility is declining. These areas, where the capacity of agro-ecosystems to continue producing food using current production methods appears most threatened, include northeast Brazil, and sections of Argentina, Bolivia, Colombia, and Paraguay.

Question 3: Determine the rate of population growth for the countries listed above,[6] and compare it to that for the United States and Western Europe. Do you see cause for concern? Explain.

The findings of significant losses of soil fertility from IFPRI analysis of nutrient depletion in Latin America and the Caribbean agree with other studies from sub-Saharan Africa (including the Sahel, mentioned above), China, South and Southeast Asia and Central America.

Soils and CO_2

While the burning of fossil fuels is thought to be the major source of anthropogenic atmospheric CO_2, soils may be a significant source as well. Soil microbes and other organisms break down the humus—the organic matter—from leaves and other living sources, and release CO_2. Scientists know that, absent human interference, soils would add an amount of CO_2 to the air in balance with that amount added to the soil from plants. But that may no longer be the case.

Atmospheric CO_2 started to increase during the nineteenth century, well before the advent of massive burning of fossil fuels. The nineteenth century growth in atmospheric CO_2 was likely derived from continent-wide conversions of woodland and prairie to agri-

[6]This information can be obtained from a number of sites, including http://www.igc.org/wri/wr-00-01/pdf/hd1n_2000.pdf, http://www.odci.gov/cia/publications/factbook/, http://www.undp.org/popin/wdtrends/wdtrends.htm.

culture in Europe and North America. The stripping of the native ground cover caused the rate of CO_2 emission from soils to increase rapidly, resulting from accelerated rate of humus decomposition. This increase in CO_2 in turn helped lead to a 0.8°F increase in temperature over the past 100 years. And this growing terrestrial temperature accelerated the decomposition of humus, a classic example of "positive feedback." Moreover, modern industrial agriculture, by cropping and erosion, removes more organic matter from the soil than it replaces, resulting in a growing imbalance between organic matter added to soils and CO_2 emitted to the atmosphere from decomposition.

And as the earth's temperature rises, each 0.9°F increase in global temperature will add an additional amount of CO_2 from the soil equal to a year's burning of fossil fuels, according to researchers at UC Irvine, thus making the problem worse.

SUMMARY: CAUSES FOR CONCERN

The primary focus of agriculture is, and must remain, the provision of an adequate quantity of decent food at affordable prices, a challenge that farmers and scientists have met with considerable success in the past. This "golden age" saw global food prices drop by 40% in real terms in the past 25 years because of crop yield increases, enlightened public policies, investments in agricultural research, and *public and environmental subsidies.*

Over the past three decades, the per capita increase in production of the world's three major cereal crops has been up by 37% for maize, by 20% for rice, and by 15% for wheat. And prices (in real terms) for these crops have dropped (down by 43% for maize, 33% for rice, and 38% for wheat). Lowering the prices of major staples directly benefits the poor, who spend a large part of their income on purchasing food.

The main reasons for these successes include:

- A continuous flow of new production technologies such as improved seeds.
- Better management practices and improved pest and disease control.
- The commercialization of farming with more efficient means of marketing outputs, a direct result of good public policies.
- The expansion of international trade that has minimized price differences between locations and seasons and fostered production patterns based on comparative advantage.

But all these advantages, as we have seen, have come at an environmental cost that has not been reflected in the price of the food.

Question 4: Identify any down side to the practices listed above. For example, are there any environmental problems associated with expansion of international trade? (Hint: Research the problem of "invasive species" in the Great Lakes and San Francisco Bay—see Issue 20.) Remember that commodity prices do not include the cost of degraded land or ecosystems. That cost has been borne by the environment.

The unprecedented scale of agricultural expansion and intensification raises two principal concerns. *First,* there is a growing concern over the vulnerability of the productive capacity of many agro-ecosystems to the stresses imposed on them by the intensification of agriculture. Can technological advances and increased inputs continue to offset the depletion of soil fertility and fresh water resources? As soil fertility reduces and water becomes scarcer, what will be the impact on food prices?

Second are the broader concerns about the negative external impacts of agricultural production that are often made worse by intensification. These negative impacts include additional stresses that agro-ecosystems can generate beyond their own boundaries but that are not properly reflected in agro-ecosystem management and production costs, nor in the prices consumers pay for food and fiber goods.

At a watershed level, these impacts include:

- Decreased river flows and groundwater levels.
- Increased soil erosion from hillside farming affecting downstream fisheries and hydraulic infrastructure (canals, dams, etc.).
- The damage to both aquatic ecosystems and human health arising from fertilizer and pesticide residues in water sources or on crops are examples of negative impact.

Finally, loss of habitat and biodiversity from putting land to agricultural uses, as well as narrowing of the genetic base and the genetic diversity of domesticated plant and animal species currently in use, are important concerns.

Among concerns expressed by the World Resources Institute[7] are the following:

- While U.S. restaurants throw away vast quantities of food left by diners at "all-you-can-eat" buffets, significant segments of the populations in the poorest countries cannot afford to purchase additional food. Thus the vast majority of new staple food supplies will need to come from domestic production in developing countries facing high population growth rates and increased threats to agricultural ecosystems.
- Most of the global agricultural production, with the exception of dairy and perishable vegetable production, still derives from intensively managed irrigated and rain-fed crop fields located far from major concentrations of population. This makes distribution of agricultural products increasingly dependent on low fuels prices and well-maintained but costly roads.
- While the net global expansion of agricultural area has been modest in recent decades, intensification has been rapid. Irrigated area grew by more than 70% over the past 30 years. While irrigated systems account for only 5.4% of agricultural land globally, they reach 35% in South Asia, 15% in East Asia, and 7% in South East Asia.

Thus, there is an urgent need for greater emphasis on monitoring land cover, soil degradation, and other indicators in order to better understand environmental effects and their relationship to agricultural productivity.

Question 5: Suggest ways that humans can address the problems posed by industrial agriculture. Recall the Issues we devoted to human population growth (1-3), for example. Cite evidence to support your answers, and be prepared to defend them against charges that they are "unrealistic."

---

[7]http://www.wri.org.

FOR FURTHER STUDY

- As you have seen, a significant fraction of the atmospheric carbon dioxide, the most common greenhouse gas, comes from burning of tropical forests, many acres of which were cleared to raise cattle. These cattle release methane, also a potent greenhouse gas. Research how raising cattle contributes to global warming.

- Consult *Global Environment Outlook—1,* by the United Nations Environment Programme[8] (available in book form or on the Internet). Research land degradation, deforestation, and water use in the Regional Perspectives section. What are the major problems? What is the outlook?

- Go to the website www.chesbay.org and research the issue of contamination of Chesapeake Bay resulting from fertilizer and manure runoff from agricultural fields. What solutions are proposed by the organization? Who will pay for them? Is it reasonable to encourage or subsidize agricultural practices which result in artificially cheap food, made artificially cheap because much of the real cost of production is borne by the environment? Do you think these practices can continue indefinitely? Explain your answers.

---

[8]United Nations Environment Programme. 1997. *Global Environment Outlook-1.* (New York: Oxford University Press). http://www.grida.no/geo1/.

AUTOMOBILES AND THE ENVIRONMENT I

KEY QUESTIONS

- What are the major automobile transportation trends in the United States?
- What are the environmental impacts of Sport Utility Vehicles (SUVs) and other light trucks?
- What are the CAFÉ standards? Are they working?

BACKGROUND

On October 1, 2000 Ford Motor Company rolled out its new $34,000+ Excursion (Figure 9-1 a and b) sport utility vehicle (SUV). Its dimensions—19 feet long by 6 feet 8 inches wide by 6 feet 4 inches tall—made it the largest SUV on the road. And, according to Truckworld Online!,[1] it still fit in a standard garage! The Limited 4x4 version came equipped with a 6.8-liter, V10, 310-horsepower engine that achieved as much as 12 miles per gallon. According to Gurminder Bedi,[2] Vice President of Ford Truck Vehicle Center, the Excursion is spacious, convenient, and versatile, "while at the same time setting a bold new standard for safety and the environment."

Let's begin our analysis by examining Bedi's environmental claim of a "bold new standard."

According to various sources,[3] about a fifth of each Excursion is made from recycled materials, including more than 1000 pounds of post-consumer recycled metal. Two-liter soda bottles, bottle caps, cotton bale wrappers, passenger car tires, and scrap battery cases all are used in Excursion components. In addition, more than 85% of the vehicle (by weight) is recyclable.

The Excursion is also classified as a low-emission vehicle (LEV).[4] This is a designation of the California Air Resources Board (ARB) based on tailpipe emissions of non-methane hydrocarbons, carbon monoxide, and oxides of nitrogen. Although LEV is a more restrictive standard than the EPA's Tier One standards that apply to all new vehicles, it is less restrictive (that is, it allows higher pollutant levels) than the ARB's Ultra

---

[1] http://www.truckworld.com/Sport-Utility/2000-excursion/excursion.html.

[2] Cited in: http://www.truckworld.com/Sport-Utility/2000-excursion/excursion.html.

[3] Including F150 online truck news (http://www.f150online.com/fordnews/11.html), SUV.com (http://www.suv.com/news/6.21.99.Ford.html), and others.

[4] California Air Resources Board. 1999. California Exhaust Emissions Standards and Test Procedures for 1988-2000 Model Passenger Cars, Light-Duty Trucks, and Medium Duty Vehicles (http://www.arb.ca.gov/msprog/levprog/cleandoc/ldvtp88.pdf).

A.

B.

FIGURE 9-1 A. Author Dan Abel next to his Ford Escort. B. The author next to a Ford Excursion. (photos by Juliana Abel/Hugh Jaahss Agency)

Low Emissions Vehicle (ULEV), Super Ultra Low Emissions Vehicle (SULEV), and Zero Emissions Vehicle (ZEV) designations.

Question 1: Based on the above information, critically assess Bedi's claim of a "bold new standard for the environment." What critical thinking standards did you apply? Describe them.

The model year 2000[5] was a watershed year for another reason as well—the first production hybrid cars, the Honda Insight and the Toyota Prius, were introduced. These vehicles use gasoline engines and electric motors to achieve mileages of nearly 70 and 50 miles per gallon, respectively. Both are also SULEV vehicles.[6]

Question 2: What is your general perception of typical U.S. car buyers? Based on this, do you think these vehicles will achieve a high share of the new car market? What additional information would you like to have to answer this question more fully?

Even with increasing gas prices, the automobile culture in the United States has never been more pervasive, and it is growing globally. We are, it is *still* said, what we drive. What are our driving trends?

Question 3: What are the unique attributes that make the personal motor vehicle so attractive?

---

[5]A model year begins on October 1 and ends on September 30.
[6]Go to the California Air Resources Board website for information: http://www.arb.ca.gov/homepage.htm.

Next, we will investigate the question: What are the costs, environmental and social, of our reliance on internal-combustion-engine-powered vehicles?

WHAT WE DRIVE

Table 9-1 shows sales data for all light trucks, SUVs, and passenger cars from 1975 to 2000. Sales for the model year 2000[7] were slightly down from 1999's record-setting pace.

Question 4: On the axes below, separately plot sales of each of the above categories for the years 1975, 1980, 1985, 1990, 1995, and 2000.

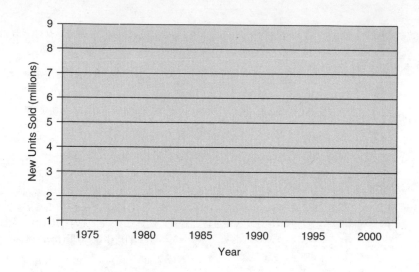

Table 9-1 ■ Sales data for various classes of vehicles.[8]

Year	Light Trucks	SUVs	Cars
1975	2,281,000	37,000[9]	8,624,000
1980	2,440,000	243,000	8,979,000
1985	4,458,000	611,000[10]	11,043,000
1990	4,548,000	931,000	9,301,000
1995	6,053,000	1,735,000	8,635,000
1996	6,519,000	2,020,000	8,527,000
1997	6,797,000	2,352,000	8,272,000
1998	7,299,000	2,695,000	8,139,000
1999		3,133,000	
2000	8,500,000	3,366,000	8,300,000

---

[7]According to the EPA, light trucks, which include pickups, SUVs, passenger vans, and minivans, are trucks or vehicles built on truck chassis with a gross vehicle weight rating of less than 10,000 lbs.
[8]Transportation Energy Data Book: Edition 20. 2000. CTA publications (http://www-cta.ornl.gov/data/tedb20/Index.html).
[9]SUV data are for 1976.
[10]SUV data are for 1984.

Question 5: Calculate the percentage of light trucks and SUVs (out of total vehicle sales) and fill in the table below.

Year	% Light Trucks	% SUVs
1975		
1980		
1985		
1990		
1995		
2000		

Question 6: Describe trends in sales of light trucks, SUVs, and passenger cars from the data in Table 9-1 and your graph.

Question 7: List as many reasons as you can why people are buying SUVs. Give the evidence you used in each case. Identify the assumptions you bring to the issue of the "suitability" of SUVs, and then critically evaluate the reasons you have just listed.

Question 8: In 1998, Ford Motor Company announced that, beginning with its 1999 model year, SUVs and minivans would reduce smog-causing emissions by 40%.[11] Currently SUVs are classified as "light trucks" and thus are allowed to exceed emissions of cars.[12] Among SUVs there is great variation in emissions, with the Chevrolet Suburban being among the "dirtiest." Should stricter emission standards for SUVs be implemented, or should reductions in emissions be voluntary, like Ford's? Justify your answer.

---

[11]Bradsher, K. 1998. Auto makers plan cuts in emissions of sports vehicles. *New York Times*, Jan. 6, 1998, p. 1.
[12]Emissions of nitrogen oxides, for example, are allowed to be 175% higher than that of the largest cars.

Question 9: Following the announcement of reduced emissions for Ford SUVs, Alexander Trotman, chairman and chief executive of Ford, said "On emissions, anyway, there will be nothing to feel apologetic about in driving a sports utility vehicle." Why did he use the word "anyway"? Did he imply that drivers of SUVs should be apologetic for something else in addition to emissions? If so, what else could there be to be apologetic about?

Question 10: According to the *New York Times* article:[13] "Thomas Stallkamp, Chrysler's then-president, said he doubted that Americans wanted cleaner vehicles enough to pay the $50 to $200 per vehicle that he estimated it would take to reduce emissions from its existing engines. 'Keeping our manufacturing costs as low as possible is important to us,' he said." What are some environmental and health impacts if emissions are not reduced? Who pays these costs? Should people who choose to drive SUVs pay these costs? Do you agree with Stallcamp's conclusion about Americans not wanting to pay for cleaner vehicles? Analyze this issue and justify your answer. Be sure to assess your reasoning, using critical thinking principles.

FUEL ECONOMY AND CAFÉ STANDARDS

At a time of turmoil in the Middle East and in response to a perceived overdependence on foreign oil, as well as a decline in fuel mileage of 1974 model year cars to an average of 12.9 miles per gallon, the Energy Policy and Conservation Act of 1975 established standards for fuel economy for passenger cars. The CAFÉ (Corporate Average Fuel Economy) standards were implemented for 1980 model year cars at 18.0 mpg. The CAFÉ standards for cars and light trucks in 2000 were 27.5 and 20.7 mpg, respectively. In practice, each automobile manufacturer produces a mix of vehicles sufficient to achieve the CAFÉ standard. If it fails, a fine of $5.50 per vehicle for each tenth of a mpg below the standard is paid. In calendar year 1999, fines exceeded $16 million.[14] Has the CAFÉ program been a success? Or has increased fuel efficiency been accomplished at too high a cost?

The automobile industry and related organizations, including the American Automobile Manufacturers Association (AAMA), and the American Iron and Steel Institute (AISI), both industry-funded groups, are opposed to higher standards, while public inter-

---

[13]Bradsher, K., *op. cit.*
[14]U.S. Department of Transportation National Highway Traffic Safety Administration Twenty-fourth Annual Report to Congress on the Automotive Fuel Economy Program www.nhtsa.dot.gov/cars/problems/studies/fullcon.

est groups generally support raising the fuel economy. Let's examine some claims made by those opposing the CAFÉ standards.[15] NOTE: Since CAFÉ standards have been frozen since 1994 by acts of Congress, we will use that date as a benchmark in our analysis.

Claim #1: CAFÉ STANDARDS HAVE NOT REDUCED OIL IMPORTS

The Data: In 1975, crude oil imports into the United States were 1,498 million barrels (1 barrel = 42 gallons). In 1994, the value was 2,565 million barrels.[16] U.S. population in 1975 was 215,973,000. United States population in 1994 was 260,651,000. Approximately 27% of the oil consumed by the United States was used in passenger cars.[17]

Question 11: Domestic oil imports increased from 1975 to 1994. Does this fact support the first claim? Have *per capita* oil imports increased over that period? If so, by what percent?

Question 12: List other factors that could have accounted for increased oil imports during that period. Justify your choices.

Claim #2: CAFÉ STANDARDS THREATEN HIGHWAY SAFETY

Question 13: What is the reasoning behind this claim, that is, how might increasing fuel economy lead to decreased safety? (Hint: One way automakers improve fuel efficiency is to remove weight from cars.)

---

[15]These claims are a compilation from the AAMA, the AISI, and others.
[16]*Statistical Abstract of the United States* 1995, Table 954.
[17]Congressional Research Service, 1995.

Question 14: Is the assumption that increasing fuel economy must be achieved only through vehicle downsizing valid? What are other ways in which increased fuel economy can be achieved?

Question 15: According to the AISI,[18] "CAFÉ has caused 2000 deaths and 20,000 injuries per year" because there are more smaller and lighter vehicles. Is this a fair and sufficiently broad and precise assessment? What actions could be taken to reduce the number of these deaths and injuries? Cite evidence to support your answer.

Claim #3: FLEET FUEL ECONOMY IS CONSUMER DRIVEN

The AAMA claims "if all consumers bought . . . 40 mpg car(s) . . . , the new car fleet average would be 40 mpg. . . . The mix of consumer vehicle purchases determines fuel economy averages."

Question 16: Is this statement accurate? Does the automobile industry's $5 billion a year in advertising play any role in influencing the type of vehicles that consumers purchase? Assess your thinking and justify your answer.

---

[18]http://www.steel.org/policy/other/cafe.asp.

Claim #4: HIGHER STANDARDS HURT THE ENVIRONMENT

According to this claim, compliance with CAFÉ standards increases the cost of new cars, thus encouraging people to buy older, more heavily polluting vehicles.

Question 17: Evaluate the validity and logic of this claim. What could be done to fairly discourage buyers from purchasing older cars? Alternatively, identify ways to reduce pollution from older vehicles.

Question 18: Does your state require vehicle emissions or maintenance testing? Would that discourage the ownership of more polluting or less efficient cars?

WHO IS USING THE GAS?

Table 9-2 shows vehicle miles traveled, fuel use, and fuel economy for passenger cars and light trucks (including pickups, minivans, and SUVs).

Table 9-2 ■ Statistics for Passenger Cars and Light Trucks, 1970–1998[19]

Year	Fuel Use—Cars (million gallons)	Fuel Use—Light Trucks (million gallons)	Fuel Economy—Cars (mpg)	Fuel Economy—Light Trucks (mpg)
1970	67,820	12,313	13.5	10.0
1975	74,140	19,081	13.9	10.5
1980	69,981	23,796	15.9	12.2
1985	71,518	27,363	17.4	14.3
1990	69,568	35,611	20.2	16.1
1995	68,072	45,605	21.1	17.3
1998	72,209	50,579	21.4	17.1

[19]Transportation Energy Data Book: Edition 20. 2000. CTA publications (http://www-cta.ornl.gov/data/tedb20/Index.html).

Question 19:   Plot fuel use for both cars and trucks on the axes below.

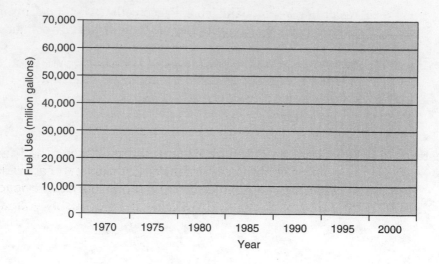

Question 20:   What trends are evident in the above graph?

For passenger cars, vehicle miles traveled (VMT) increased from 0.9×10^{12} miles in 1970 to 1.5×10^{12} in 1998. For light trucks the corresponding figures are 1.2×10^{11} and 8.7×10^{11}.

Question 21:   Calculate the average annual percentage increase in vehicle miles traveled for both cars and light trucks for the period 1970–1998. Use the following equation (which is a rearrangement of compound growth equation; see *Using Math in Environmental Issues,* pages 14–15):

$$k = (1/t)\ln(N/N_0)$$

where k = the percentage increase per unit time; t = the number of years; N = the variable (i.e., vehicle miles traveled) at the end of the given time; and N_0 = the variable at time 0 (i.e., the beginning of the period we're interested in).

Question 22:   Interpret your findings in Question 20. What could plausibly account for the rates you calculated?

Question 23:   Summarize what you have learned in this issue.

Question 24:   Refer back to Question 6. One of the reasons that people are attracted to SUVs and larger vehicles in general is that they can carry a lot of "stuff." Is this reason enough to own a vehicle that disproportionately impacts society? In other words, do we need all the stuff we carry around? Justify your answer.

FOR FURTHER STUDY

- We discard over 250 million tires a year. Cadmium and other heavy metals from tire wear, for example, pose a significant threat to estuaries like San Francisco Bay and Chesapeake Bay. Tire dust causes human respiratory problems.[20] Evaluate the environmental impact of tire use and discards. How can we reduce tire impact while increasing personal motor vehicle use?
- For an authoritative and readable examination of the impact of cars, read *Asphalt Nation: How the Automobile Took Over America and How We Can Take It Back* by Jane Holtz Kay (1997, University of California Press, Berkeley, CA).
- Obtain information on current gasoline consumption from the Energy Information Administration[21] (EIA). Is it up or down from previous years? Explain your observation.
- Research fuel economy of new cars. For 1998 models, for example, nine out of every ten get fewer than 30 miles per gallon.

---

[20]*Rachel's Environment and Health Weekly.* 1995. Tire Dust. #439.
(http://www.rachel.org/bulletin/index.cfm?St=2).
[21]http://dir.yahoo.com/Government/U.S__Government/Executive_Branch/Departments_and_Agencies/Department_of_Energy__DOE_/Energy_Information_Administration/.

AUTOMOBILES AND THE ENVIRONMENT II: GLOBAL TRENDS

KEY QUESTIONS

- What are the major automobile transportation trends outside of the United States?
- What are the environmental impacts of cars and light trucks?
- Are there sustainable transportation alternatives?

THE ENVIRONMENTAL COST OF MOTOR VEHICLES

The environmental and social impact of our reliance on internal-combustion-engine-powered automobiles and light trucks in general, is profound. In the United States, these impacts include[1]:

- 41,000+ deaths a year in automobile crashes.[2]
- 2,000,000+ serious injuries annually; healthcare costs associated with these injuries.
- Annual interest cost over $10 billion to finance the purchase of passenger cars.[3]
- Hundreds of millions of vertebrates killed each year.
- Air and water pollution (see below), including a major contribution to what the Natural Resources Defense Council calls "poison runoff." Poison runoff includes cadmium, zinc, and other heavy metals from tire wear, hydrocarbons from fuel and grease.[4]
- $3 billion a year spent for catalytic converters on new vehicles.
- Accelerated urban sprawl as roads are paved.
- Other impacts of roads (road salt, asphalt, fragmentation of habitat, pollution associated with road construction, increased risk of flooding, etc.).
- Increased dependence on foreign oil.
- Noise pollution.

Burning fossil fuels produces gases. Under the "cleanest" conditions these would be H_2O and CO_2. However, since air is used to burn the fuel (e.g., in a car engine) and since

---

[1]For details, consult websites of these organizations AAA, EPA, NRDC, Wildlands CPR, and others.
[2]Insurance Institute for Highway Safety data for 1999 (http://www.hwysafety.org/safety_facts/safety.htm).
[3]1998 data, from http://www-cta.ornl.gov/data/tedb20/Chapter_5.pdf.
[4]Poison runoff also originates from nonpoint sources (like golf courses; as opposed to point sources such as sewer pipes or factory smokestacks).

air is 70% nitrogen (N), nitrous oxides (NOx) are produced. Furthermore, since sulfur is present in most fuels, particularly coal, diesel fuel (which can contain up to 500 ppm legally as of 2000), and gasoline (up to 330 ppm), oxides of sulfur (SOx) may be produced when the fuel is burned. Carbon monoxide (CO) is usually produced as well; when properly maintained, catalytic converters change most of the CO to CO_2. And, if some of the liquid fuel (diesel or gasoline) is not completely burned (which is typical of engines before they are warm), various volatile organic compounds (VOCs) are produced. VOCs and other gases react with sunlight to produce ozone (O_3). Burning diesel fuel also produces soot (particulate matter), which has been implicated in respiratory diseases, including lung cancer, emphysema, and asthma.

So, the emissions from motor vehicles may include NOx, SO_x, CO, VOCs, ozone, H_2O, and CO_2. Let us review the toxic effects of these emissions. NOx and SOx react with moisture in the air and result in acid precipitation. Furthermore, recent research has implicated SOx emissions in the production of methyl mercury by bacteria in soils and wetlands (see Issue 26: Methyl mercury is the most toxic form of this metal.) CO is harmful to organisms and can be lethal at high concentrations. VOCs and fine particles are carcinogens (cancer-causing agents). Although CO_2 is not a pollutant, it is a greenhouse gas; its impact is discussed in Issue 18. Ozone can harm plants such as tomatoes and corn; low concentrations can cause eye irritation, and high concentrations can damage animal respiratory systems.

AUTOMOTIVE USE IN CHINA

In the United States, drivers are abandoning smaller, fuel-efficient passenger vehicles in favor of larger cars, minivans and SUVs.[5] Are the trends similar in other countries? In this issue we will consider the global environmental impacts of increased automobile emissions resulting from an exponential growth of automobile use in other countries. We will focus first on China.

Along with other developing countries such as India and Indonesia, China is a country in which motor vehicle use will probably increase substantially over the next several decades. With 1.3 billion people, China has the potential to be a global economic superpower in the twenty-first century. It already has an enormous environmental footprint, as we shall see.

First, some background: A 1994 article in the *Washington Post* entitled "Dreams on Wheels"[6] described a car purchase by Wang Xian and his wife, members of the small but growing Chinese middle class. Mr. Wang was one of only 350,000 Chinese to purchase a car in 1994, but the official newspaper, *China Daily,* enthused over a domestic market consisting of "about 300 million potential car owners."

Approximately 1.2 million personal motor vehicles (PMVs, that is minivans, SUVs, cars, and pickup trucks) were sold in China in 2000.[7] Automotive Resources Asia[8] projects sales of 1.976 million per year by 2005. Let us consider the impact of adding millions of cars per year to roads (Figure 10-1) in China and adding a total of 300 million eventually.

---

[5]The newest trend is crossover vehicles, hybrids between passenger cars and SUVs.
[6]Mufson, S. Dreams on Wheels. *Washington Post,* December 28, 1994.
[7]Smith, Craig S. The race begins to build a small car for China. *New York Times,* October 24, 2000.
[8]http://www.auto-resources-asia.com/m_mktoutlook_china.php.

FIGURE 10-1 Authorities bar cyclists from a street in Beijing, reserving it solely for motor vehicles. (Elisabeth Rosenthal/The New York Times)

Question 1: For the model year 2000, China's new vehicle market was 23,686 SUVs and 626,300 cars.[9] For 2005, these are projected to be 39,000 and 976,600 respectively. Refer to Issue 9 and compare the vehicle mix and growth rate for cars and SUVs for China to those for the United States. Comment on your findings.

Question 2: If vehicle sales were 350,000 in 1994 and rise to 1.976 million by 2005, what would be the annual growth rate in sales per year? (HINT: Use the formula $k = (1/t)\ln(N/N_o)$, introduced in *Using Math in Environmental Issues,* pages 14–15.)

Question 3: At this rate of increase, how long would it take China to put 300 million vehicles on the road? Use as a starting point 1.976 million PMVs in 2005. (HINT: Use $t = (1/k)\ln(N/N_o)$, introduced on page 15.)

---

[9]Ibid. Other categories ("minis," buses, other trucks, etc.) have been ignored for this calculation.

Question 4: Assume each car averages 24 mpg and is driven 10,000 miles each year. How much gas, on average, would each car use per year? Express your answer in gallons and liters. (Some analysts believe the Chinese mpg average will be closer to 30 mpg. Keep in mind the *assumptions* under which you are doing this analysis.)

Question 5: How much gasoline would be required for 300 million cars annually? What is the daily requirement in barrels per day (1 barrel = 42 U.S. gal)?

Question 6: The United States is the world's largest consumer of gasoline, with the present U.S. demand at around 8 million barrels per day (1 barrel = 42 gallons). Compare this figure with the projected Chinese demand for gasoline.

During 2000, global oil production was around 78 million barrels per day but had experienced considerable fluctuation over the past decade, due to severe price swings (see Issue 14). But by 20__ (insert your answer from Question 2 in the preceding blank and from Question 4 in the following space), the world would have to produce at least an additional __ million barrels per day to satisfy the Chinese demand. Is this possible? If so, what are the economic implications?

Question 7:   What do you think will happen to the price of oil if global demand rises as you forecast to meet Chinese demand? (A good place to find a projected future price of oil is the U.S. Energy Information Administration's Annual Energy Report.[10]) Assume that the price of oil will increase and discuss the impact of this situation on poor countries. (List any other assumptions you will have to make to discuss these questions.)

Now let's consider the environmental impact of these 300 million cars.

Question 8:   Assuming 1 g NOx per mile driven for vehicles with minimal pollution control devices, how many metric tonnes of NOx would be generated by 300 million vehicles driven a modest 10,000 miles per year?

Now compare this number to the total NOx produced in the United States, which was 23.6 million tons (not *tonnes*) from all sources, not just cars, in 1999.[11]

CRITICAL THINKING QUESTIONS

Question 9:   Following is a quote from Samuelson[12]:

> . . . [T]he central ambition of postwar [American] society has been to create ever expanding prosperity here and abroad. This is not because Americans worship materialism, although it sometimes seems that way, but because prosperity has seemed to be the path to higher goals. At home, it would end poverty and the associated ills of crime, slums, and racial conflict. It would underwrite more generous government to support the elderly and the disabled. It would expand personal choice and freedom. Relieved of material wants, Americans would have more opportunity to express their individuality and enjoy themselves. Abroad, global prosperity

[10]http://www.eia.doe.gov/oiaf/aeo/.
[11]U.S. Bureau of the Census. 2000. *Statistical Abstract of the U.S.: 1999*
(http://www.census.gov/prod/99pubs/99statab/sec06.pdf).
[12]Samuelson, R.J. 1995. *The Good Life and Its Discontents* (New York: Times Books).

would . . . spread democracy, and solidify U.S. global leadership. The United States would be the world's role model; our democracy would be the most admired, our economy the wealthiest. Other countries would emulate our political institutions and management practices. Our domestic and foreign ambitions have been spiritual twins. Both rested on a deep faith in the power of prosperity to improve the human condition.

Discuss where you agree and disagree with Samuelson's picture of U.S. postwar goals. Do you think Americans "worship materialism"? How does this question relate to the carrying capacity concerns raised in Issue 3? Discuss.

Question 10: Do you think prosperity creates or helps solve environmental problems? Discuss.

Question 11: What is your evaluation of the overall impact of a consumerist society in China, based on the U.S. model? (What is a "consumerist" society, anyway?) Do you think a consumerist society in China would be conducive to political stability in Asia? In the world? Why or why not? Again, recall the discussion of carrying capacity from Issue 3 when answering this question.

Question 12: Describe a scenario in which the United States would be drawn into potential conflicts in Southeast Asia, based on China's demand for petroleum or China's air and water emissions. Do we have any treaty responsibilities with countries in the region that might be drawn into confrontations with China over environmental pollution?

Question 13: Some options to diesel- or gas-powered vehicles are natural gas, ethanol or methanol, fuel cells, "hybrid engines," and electric cars. There is also the possibility that the Chinese could simply "leapfrog" beyond a fossil-fuel based transport system by requiring production of vehicles that use a fuel other than petroleum. Do you think U.S. automobile makers are poised to take advantage of such a new market if it were to arise? In other words, do U.S. automobile makers produce such vehicles at present? (For more information, contact Ford, General Motors, DaimlerChrysler, Honda, or Toyota for a copy of its annual report, or call its public relations department [PR departments usually have toll-free numbers] and find out how much, if anything, it is investing in alternatively fueled or hybrid automobiles.)

GLOBAL TRANSPORTATION TRENDS

The potential for growth of the motor vehicle fleet worldwide is very high. By the year 2025, a projected 1 billion vehicles will be found on the world's roads.[13] While per capita ownership of motor vehicles is low in developing countries, as we have seen in the case of China, it is on the rise. In China and India there are between 7 and 8 vehicles per 1000 people. These countries still have a long way to go to reach the level of ownership in the United States, about 750 motor vehicles per 1000 persons.[14]

There are some hopeful signs (e.g., new mass transit in Equador, banning of old buses in Delhi, the phase-out of leaded gas in Manila) that the increase in motor vehicles in de-

---

[13]http://www.wri.org/wri/trends/autos2.html.
[14]American Automobile Manufacturers Association (AAMA).1996 *Motor Vehicle Facts and Figures 1996* (Washington, DC: AAMA), pp. 44–47.

veloping countries will include environmental regulations, sensitivity to alternative forms of transportation, and sustainably designed vehicles.[15]

FOR FURTHER STUDY

- Where could "alternative fuel" for 3×10^8 cars come from? Coal? Nuclear fission? Solar power cells? Research this issue, especially the Chinese coal industry and the potential to produce liquid fuel from coal. List and discuss the environmental implications.

- Based on 300 million motor vehicles, the Chinese would have to dispose of or recycle 300 to 600 million tires per year. Research the problem posed by tire disposal in the United States and discuss.

- Study the prevailing wind direction in Southeast Asia. You can find this in any introductory meteorology text. These winds are called monsoons. Discuss the distinctive attributes of monsoon winds. Plot the wind directions on a map of Southeast Asia.

- In terms of political relationships, China and Vietnam have been at odds for hundreds of years and fought a brief war as recently as the late 1970s. Put that fact together with the well-established environmental relationship between NOx emissions, acid rain, and resultant damage downwind to human health, structures, forests, and crops. Therefore, based on your study of the monsoons, do you think the massive NOx pollution that would be generated by Chinese automobiles would have an adverse impact on Vietnam (e.g., what would be the impact of an additional 6 million tonnes of NOx on Southeast Asian agriculture, forests, reefs, water quality, air quality, and healthcare costs?)?

- For more information on China's environmental problems, read *Our Real China Problem*, by Mark Hertsgaard,[16] or visit Ward Communication's website for a summary of China's "Family Car" Strategy for the 21st Century.[17]

- Research the impact of fossil fuel emissions on human health in China at http://www.wri.org/wri/wr-98-99/prc-air.htm. Summarize your findings.

---

[15]http://www.itdp.org/.

[16]Hertsgaard, M. Our Real China Problem. *The Atlantic Monthly,* November, 1997, pp. 96–114.

[17]Ward's Communications: China's "family car" strategy for the 21st Century [summary], available (with ordering information) at: http://www.wardsauto.com/specrep/china.htm.

WASTE, PACKAGING, AND SUBSIDIES:
THE ALL-ALUMINUM CAN

KEY QUESTIONS

- What is the environmental impact of producing and discarding aluminum cans?
- What is the nature and significance of subsidies in aluminum production?

BACKGROUND

In 1997, Reuters news agency reported that the Coca-Cola Company test-marketed a "new, embossed" aluminum can for its Sprite and Coke beverages:

> The embossed can includes an indentation where cans for the colorless soda now have bubbles and a tear-like design that Coke executives call a "silver streak." Coke said the can has high registration graphics that can be seen more clearly. [A Coca-Cola Company spokesman said] that the embossed can is another indication of the company's new strategy of using packaging to increase perceived value for customers, and "it gives this can a really incredible appearance, a wonderful feel." Consumers [have said] they like the look, the feel, and in some cases, say it makes the product taste better.

The food and beverage distribution industry has long recognized the importance of packaging as a marketing tool, but packaging can impose considerable environmental cost. Furthermore, the industry amply illustrates the subsidies, energy and otherwise, built into our economic system. As you just read, the appearance of the package (an embossed can in this case) can sometimes make the product taste better! Let us consider the example of the all-aluminum can.

ALUMINUM MINING AND REFINING

Aluminum (symbol Al) is one of the most common elements in the Earth's crust, but ores of aluminum are rare. The most common ore of aluminum is bauxite. Bauxite was first discovered in the French district of Les Baux in 1821, from which it took its name. Bauxite is found in extensive deposits in many countries, but mainly in the tropics and subtropics, including Australia, Surinam, Jamaica, and India (Figure 11-1). This is because bauxite is typically formed where extreme weathering conditions exist, brought about by high annual temperatures and abundant precipitation. These conditions leach out all the easily soluble elements from surface rock and sediment, leaving behind only the most insoluble elements and compounds. Bauxite, along with iron oxides and hydroxides, is the most common substance produced by such extreme weathering. Even though bauxite is a rich ore, the aluminum metal is tightly locked into the mineral structure and is not easily extracted.

FIGURE 11-1 Bauxite mine. (Yann Arthus-Bertrand/CORBIS)

PROCESSING BAUXITE ORE

First the bauxite must be crushed, then processed into a white powder called alumina (Al_2O_3). During processing, four tonnes of bauxite ore yield two tonnes of alumina and two tonnes of waste. The two tonnes of alumina further reduce to one tonne of Al metal.[1]

Question 1:   How much waste is produced per tonne of Al refined?

The alumina is then dissolved in molten cryolite (Na_3AlF_6) to make the solution conduct electricity. This molten solution is placed in a large vat and a powerful electric current is passed through it. The current separates the Al from the solution and molten aluminum metal sinks to the bottom of the vat, where it is drawn off. A by-product of the reaction, HF, is extremely toxic and is released as a gas. (A partnership with the U.S. Environmental Protection Agency will help producers reduce HF by 50% by 2010.) CO_2, a "greenhouse" gas, may also be emitted in the process.

A great deal of electricity is needed to produce molten aluminum, the form that is then made into foil, sheet metal, and so on (see below). In fact, this very capital- and

---

[1]http://www.alcan.com/.

energy-intensive process accounts for 2% to 3% of the electricity used in the United States every year! There is a significant potential for improvements in the efficiency of the process. Theoretically, the current but inefficient "Hall-Heroult" reduction process could be retrofitted with new technology cathodes, with potential energy savings nationwide of 1500 megawatts (the electricity produced by three typical coal-fired power plants).

Refiners sometimes try to solve the energy supply problem by building their own hydroelectric dams. In Canada, aluminum producer Alcan owns hydroelectric dams with a total installed capacity of 3,583 megawatts (enough to supply a city bigger than San Francisco), of which 2,740 are classified as firm power capacity and supply the needs of Canadian smelters. Alcan smelters in Scotland and Brazil operate their own hydropower dams but must purchase some additional power while the company's smelters in Switzerland and Norway buy electricity from others.[2]

ALUMINUM CANS

Cans are made from thin sheets of refined aluminum. The efficiency of can manufacture has been increasing. Here are statistics from the Aluminum Association:

Year	# of Cans per lb of Al
1972	21.75
1973	22.25
1974	22.70
1975	23.00
1976	23.30
1977	23.47
1978	23.65
1979	23.69
1980	24.23
1981	24.45
1982	25.21
1983	25.70
1984	26.00
1985	26.60
1986	27.00
1987	27.40
1988	28.25
1989	29.30
1990	28.43
1991	28.87
1992	29.29
1993	29.51
1994	30.13
1995	31.07
1996	31.92
1997	32.52
1998	33.04
1999	33.10

[2]Ibid.

Question 2: Graph the relationship on the axes below. Did the slope of the line change?

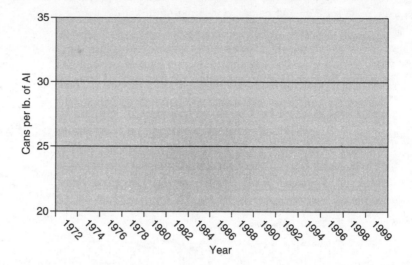

Question 3: Project the value to 2010 and 2020, assuming cans can continue to get lighter without compromising the integrity of the can. (There are a variety of ways to do this: You can either extrapolate from your graph, calculate the slope and project using it, or calculate the most recent growth rate and use $N = N_0 \times e^{kt}$.) How much did each can weigh in 1972? In 1982? In 1992? In 1999? How much would each can weigh in 2010? In 2020?

RECYCLING RATES

We next want you to analyze the recycling rate of cans. Al melts at 660°C, so it's easy to recycle. Al cans are very valuable, since it takes only 5% of the energy to make new cans from old cans than it takes to make new cans from bauxite, and of course, a great deal of waste is avoided as well (see above). For example, for each pound of Al produced by recycling, 3 pounds of waste are not generated.

Consider these data:

Year	Recycling Rate (%)
1973	15.2
1974	17.5
1975	26.9
1976	24.9
1977	26.4
1978	27.4
1979	25.7

Question 4: Plot these data on the axes below, then project the recycling rate for 1999 as above.

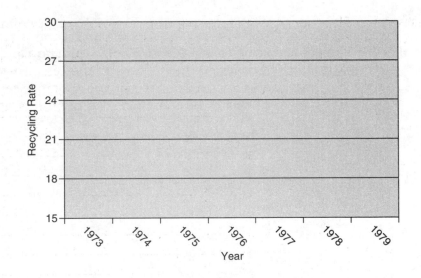

Here are the rest of the data:

Year	Recycling Rate (%)
1980	37.3
1981	53.2
1982	55.5
1983	52.9
1984	52.8
1985	51.0
1986	48.7
1987	50.6
1988	54.6
1989	60.8
1990	63.6
1991	62.4
1992	67.9
1993	63.1
1994	65.4
1995	62.2
1996	63.5
1997	66.5
1998	62.8
1999	62.5

Question 5: Plot the recycling rate changes from 1972 to 1999 on the axes below. Did the recycling rate you projected from the 1972–1979 data agree with the actual one? Did this reinforce or challenge your faith in the accuracy of projections? Why?

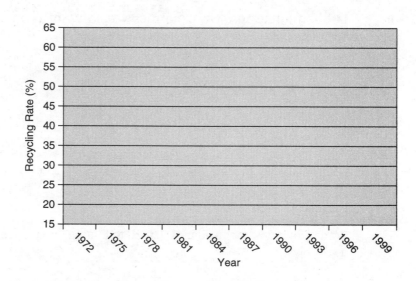

Question 6: Are you able to project a "reliable" recycling rate for 2010? 2020? Explain. Why do you think the recycling rate seems "stuck" in the low 60s? List the factors you believe can affect the recycling rate for Al cans.

SUBSIDIES AND ALUMINUM PRODUCTION

The electricity to extract the aluminum is only part of the cost to produce aluminum cans. Other costs include the extraction and shipping of the ore, the cost to restore the surface environment after mining is completed (or if restoration is not carried out, this cost is simply "dumped" onto the local environment and residents), the energy to ship the ore or concentrate to the refinery, and the cost to transport and distribute the cans. Economists define "externalities" as a cost that is not included in the price users are charged for a commodity. Subsidies are a type of externality. According to the Worldwatch Institute, many activities in industrial countries come with two price tags, one obvious, one concealed. In Germany, for example, generating electricity from coal costs a utility 6 cents per kilowatt-hour, but it costs the population at large 2 cents more, if one factors in the disease and death caused by air pollution.[3]

Subsidies tend to distort markets, since subsidies encourage use of a commodity in excess of the level that the commodity would be used if all the costs to produce the commodity were included in the price. Countries that wish to encourage mining activity often subsidize the mining process by allotting grants, tax refunds, and rebates, and other inducements, legal and otherwise, to mining companies.

Question 7: List some reasons why countries would wish to encourage mining by awarding subsidies.

Some examples of subsidies awarded to mining companies (including bauxite mining) are as follows:

■ The Australian government has awarded mining companies a 31 cent per liter rebate on diesel fuel (farmers get a 34 cent rebate). Even with the rebate, diesel fuel costs miners up to 7 cents more per liter in Australia than in competing countries,

---

[3]http://www.worldwatch.org/pubs/paper/134a.html.

such as Indonesia, Chile, and Canada. In other words, these latter countries subsidize the mining industry more than Australia[4]!

■ States in India underwrite power plant costs for miners, provide sales tax exemptions, exemptions from duty on electricity and diesel fuel, and investment subsidies for a portion of the capital cost of the entire project.[5]

Worldwide, the Worldwatch Institute[6] calculates that "environmentally damaging subsidies (of all kinds, not solely in mining) cost taxpayers and consumers more than $500 billion each year."

ALUMINUM REFINING AND ELECTRICITY COSTS

Refining aluminum from ore is extremely energy intensive, so most aluminum refiners either operate their own power dams as you saw above, or are concentrated in areas where electric rates are abnormally low. Such an area is the Pacific Northwest, where a federal agency established in 1937 by the Roosevelt administration, the Bonneville Power Authority (BPA), operates thirty power dams on the Snake and Columbia Rivers (Figure 11-2). Under legislation that created the BPA, the agency must first serve public power agencies such as municipal utilities, as well as rural cooperatives and a half-dozen government agencies. Some of what is left must go to investor-owned utilities. What is left over has been customarily sold to private industries, such as the aluminum companies. Where there was once plenty of leftover power, population growth in the Northwest and a booming high-tech economy led by Microsoft and Boeing have resulted in a power shortage. By 2000, Bonneville was forced to go into the market and purchase about 1000 megawatts of capacity.

BPA, as of early 2001, is putting the finishing touches on electric power supply contracts that will run from October 31, 2001 through September 30, 2006. It appears that aluminum smelters, who built their mills in the region during and after World War II because of the abundance of power available from Bonneville's dams, will be hardest hit, as the federal agency no longer has power to spare and must go into the market to buy electricity in all but the wettest of years.

The BPA charges aluminum refiners a little under 3 cents per kilowatt-hour (kWh), which is one-half the average rate paid by industrial consumers in the United States. Total subsidies provided by the BPA have been estimated at $1 billion per year,[7] and the aluminum refiners' portion of this has been calculated at $200 million per year. However, in 2000 BPA experienced an unprecedented jump in demand for electricity. "Our customers are asking for possibly 1400 average megawatts more than we anticipated," a BPA spokesperson said. Skyrocketing demand for power in the region is putting the aluminum refiners' favorable power rates at jeopardy. The aluminum companies would like 3000 megawatts, but won't get it. In the most recent contract, they received 2000 megawatts and that will be cut to 1500 megawatts under the plan allocation now being readied for signature.

Will increased demand put an end to subsidies for aluminum refiners? Read the following report from BPA and answer the following questions.[8]

---

[4]Chamber of Mines and Energy of Western Australia (http://www.mineralswa.asn.au/).
[5]Indiaserver http://www.indiaserver.com/biz/dbi/MEA527.html.
[6]http://www.worldwatch.org.
[7]Georgia, P. 1995. *Competitive Enterprise Institute (CEI) Position paper on Subsidies*. (Washington, DC: The Institute).
[8]http://www.bpa.gov/corporate/kcc/jl/00jl/jl1200x.pdf.

FIGURE 11-2   John Day hydroelectric dam and power plant next to Martin Marietta aluminum refinery, Washington state. (©James L. Amos/CORBIS)

PORTLAND, Ore., Nov. 8, 2000—Customer demand under new 10-year wholesale power contracts with the Bonneville Power Administration is so strong that the agency must purchase power on the market to augment its supply. "To recover the cost of these purchases, the agency proposes to tack a 15 percent charge onto wholesale rates that go into effect Oct. 1, 2001," said Paul Norman, BPA senior vice president.

Even with an additional charge, BPA's wholesale rates are still well below wholesale market prices currently forecast for the next five years on the West Coast. New agreements pushed the agency's total firm energy load (the amount of energy BPA must supply) up to 11,000 megawatts. That's nearly 3000 more than the federal Columbia River Power System can generate on a reliable basis.

"Costs in the deregulated wholesale power market have become very volatile," Norman explained. "Recent steep increases and an apparent upward trend in market prices have prompted utilities formerly purchasing power in the market to bring their business back to BPA.

In 2000, BPA filed proposed power rates with the Federal Energy Regulatory Commission (FERC) for the October 2001 to October 2006 period. The rate for public agency customers and the residential and farm customers of investor owned utilities was set at about 2.2 cents a kilowatt hour, almost unchanged from rates in effect since 1996. BPA needs this adjustment to ensure timely repayment of obligations to the U.S. Treasury for investment in the Federal Columbia River Power System. This is essential to maintaining the region's low, cost-based rates. BPA's proposal includes the ability to invoke additional price increases later in the five-year period if needed to maintain financial stability and meet all of its obligations.

Question 8:   What is the imbalance between supply and demand forecast to be under the new agreements?

Question 9: How does BPA propose to provide this power?

Question 10: Does providing industry, or residents, electricity at very low rates encourage conservation or efficiency? Why or why not?

Here is the aluminum refiners' response to BPA's new contract offer.

> Nov. 2, 2000—Aluminum companies in the Pacific Northwest signed their power supply contracts with the Bonneville Power Administration before the October 31 deadline, but did so with considerable grumbling and said yesterday that the prices they will have to pay for electricity will prevent them from going back into full production.
>
> Bonneville said that five aluminum companies operating ten smelters, two chemical companies and one paper mill signed contracts that run from Oct. 1, 2001 through Sept. 30, 2006. The taxpayer-owned utility will allot 1486 megawatts of power among those customers at a charge of $23.50 per megawatt-hour. It isn't enough, according to some.
>
> Brett Wilcox, owner of Golden Northwest Aluminum, said the smelters, attracted to the Pacific Northwest more than 50 years ago by plentiful and inexpensive electricity, need about 3000 megawatts to get back into full production. Wilcox said one of his two smelters, a Goldendale, WA, facility, was negotiating with independent power producer Goldendale Energy Inc. for development of a new 248 megawatt plant that would allow the aluminum refinery to resume full output. The gas-fired, combustion turbine plant is expected to begin commercial operation in about two years.

Question 11: What rate did BPA offer the refiners? Express it in cents per kilowatt-hour and compare it to the average national industrial rate of 6 cents per kWh.

THE FREE INTERNET ISN'T FREE

In case you wonder where the new demand for power in the Pacific Northwest is coming from, read this report.[9]

> Oct. 24, 2000—A report in the Ziff Davis on-line *Interactive Week* trade publication for Internet businesses has laid the blame for power shortages across the U.S. and in Europe on the In-

---

[9]http://www.energyonline.com/Restructuring/news_reports/news/1024tech.html.

ternet itself. According to the report, some of the huge "server hubs" that collect data and send them on over the Internet represent a load of about 35 megawatts. There are a lot of them and more are being added every day. Operators of the server hubs are now being told that they will have to wait 18 months or more for power, news that is met with disbelief in an industry where yesterday is ancient history. Some utilities are telling the dot-coms that power simply won't be available in the foreseeable future while others are scrambling to purchase firm power for resale to the Internet companies.

The report said the crunch is most acute in New Economy centers, such as the San Francisco Bay Area and Seattle, but is being felt in other parts of the U.S. and in Europe, where electric power is in even shorter supply. On October 9, the Seattle City Council enacted a law authorizing the city's municipal utility, Seattle City Light, to charge a higher rate to new industrial customers seeking large quantities of power. The new rate level hasn't yet been established, but utility spokesman Dan Williams said the premium price Seattle City Light would have to pay for the extra power would be passed on to the new customers.

Seattle City Light expects a few new data centers to increase its load by 250 megawatts over the next two years. The utility will also put into effect connection charges that will enable it to recover on the spot any investments in facilities needed to serve new customers, up to a new substation or transmission or distribution lines.

Puget Sound Energy Corp., the investor-owned utility that serves much of western Washington, said that since January 2000 Internet companies have asked for 700 megawatts of power. Steve Secrist, a spokesman for the company, said Puget Power wants to charge those companies up front for any infrastructure investment they require, just like Seattle City Light, "to make sure these new customers are not putting a strain on (our) other customers."

Question 12: How much new power did Internet companies request from Puget Sound Energy between January and October 2000?

Question 13: How much did Seattle City Light expect its demand to rise from new data centers?

Here is what the Sierra Club says about BPA's rate structure, and its subsidies to industry.[10]

> We Northwesterners treasure our lands—the rainforests, the mountains and arid lands of the inland empire. We also treasure our own little subsidy, the lowest electrical rates in the United States. Why are we so lucky? Because we have one of the greatest energy systems in the world, the Columbia River.

---

[10]http://www.sierraclub.org/chapters/wa/crest/crest295/electric.html.

The dams along this great river have spawned entire industries: aluminum smelting; vast desert orchards irrigated with water pumped at incredibly low cost; barging consortiums enjoying free transport through extensive and expensive locks; and, among the most controversial, capital-intensive fish hatcheries that employ a small army of technicians. The other great benefit is accrued by the rest of us. We pay a tiny tariff for our residential power as compared to the rest of America's citizens.

As residents of the Columbia watershed we owe, via the BPA, $6.7 billion to the U.S. Treasury for construction of the dams and power lines. But even though we willingly accept our bounteous good fortune, we have routinely defaulted on this debt and will do so again this year. Worse, through BPA policy and our fish and wildlife departments, we have destroyed one of the largest free natural resources in the entire world and replaced it with a costly, ineffective and often destructive copy called "hatcheries." And we cling to the notion that higher rates will destroy our economy although citizens elsewhere in the U.S. have to cope with much higher energy costs in order to do business.

Our rates are low because we have refused to look at our costs. We have deferred everything, including debt and long-term environmental costs, to our children, yet we are kicking and screaming all the way, saying we deserve these low rates. When do subsidies for one group become welfare for another?

Developing a happy customer base is the goal of BPA. If the aluminum industry ceased to exist, the power sales lost to BPA would entail layoffs and capital reductions within BPA and probably the Corps of Engineers. In order to prevent this, BPA wants to maintain its customer base regardless of the environmental problems that it creates. . . . Until we, as customers, realize our roles in artificially maintaining low rates and demand a system that pays for itself, we are party to the low rate subsidy cycle. Altering the months of peak hydropower production, changing Corps dam-construction plans and reducing subsidies to aluminum manufacturers and irrigators would cost the average individual ratepayer $1.20 to $2.00 per month—leaving us with, guess what, the lowest rates in the nation.

Question 14: Assess using critical thinking principles what you believe to be the Sierra Club's position and its assumptions regarding the Pacific Northwest's artificially low electricity costs. How do they contrast with BPA's position? Who would lose if the aluminum industry "ceased to exist?"

ELECTRICITY AND ALUMINUM CANS

According to industry statistics, in 1999 production of one aluminum can required 0.35 kWh of electricity.

Question 15: What is your cost for electricity per kWh? (Check your last bill, ask your parents or your college's physical plant, or ask your local utility what it charges for residential users. Nationwide, the average cost to individuals is around 8 cents per kWh.)

Question 16: At this price per kWh, how much would an aluminum can cost you to produce?

Now factor in raw materials ($25 billion in 1997), labor ($3.7 billion), and other costs, which amount to 14.5X what the industry paid for energy in 1997. Is it fair to assume that this would add at least an equal amount to the cost of each can? Why or why not?

Let us assume that it is a fair number, which would mean the total cost to produce an aluminum can would be 2.8 cents + 2.8 cents = 5.6 cents, and this does not include any externalities.

Question 17: Check the sale price of a six-pack of a generic or store-brand soft drink in aluminum cans at your local supermarket. Write it below.

Question 18: How "can" this be? (That is, explain how the cost to produce an aluminum can, using typical electricity rates paid by residential users, would be an appreciable portion of the selling price of many canned drinks.)

In 1999, 102.2 billion aluminum cans were produced in the United States and 37.5% were discarded (i.e., not recycled).

Question 19: How many cans were discarded? At 0.35 kWh per can how many kWh were thrown away with the aluminum?

WHACKER MADNESS? THE PROLIFERATION OF TURFGRASS

KEY QUESTIONS

- What is the economic impact of turfgrass and lawns?
- What are the positive and negative environmental impacts of turf grass proliferation?

BACKGROUND

A reader wrote to Ann Landers[1]:

> Dear Ann:
>   Why does "Keeping Up With the Joneses" think he is doing his neighbor a favor mowing his lawn? And you called him generous? Try annoying. . . . We bought a home in the woods thinking we would get away from the weekend Lawn Rangers trail. Our neighbors have this golf course mentality, which is positively maddening. One in particular spends every waking minute using some kind of monstrous motor-driven vehicle to maintain his lawn. If it isn't the lawn tractor, its the weed whacker, the leaf blower, the power mulcher, the lawn vac or a chain saw to get rid of the trees that block his sun and drop leaves or pine needles. And then he runs his water pumps for the lawn sprinklers. Enough already!
>   It would be really nice to lie in the hammock and have some peace and quiet. No such luck. On nice days in the summer, we are forced to stay inside with the windows shut and the air conditioner turned on just to hear each other talk.
>   We spend a minimal amount of time mowing down the weeds, and we encourage the moss to grow beneath the pine trees because there are better things to do in life than mow lawns and contribute to noise pollution. They probably refer to us as—
>   The Schlocky Neighbors in Knowlton, Wisc.

LAND-USE CHANGES IN NORTH AMERICA

Before the arrival of Europeans, most of the eastern United States was hardwood forest, but changes in land use during the ensuing 400 years have been profound, as documented in sediment cores taken from Chesapeake Bay[2] (see Issue 19). The first major change in-

---

[1]December 19, 1997.
[2]Cooper S.R., and G.S. Brush. 1991. Long-term history of Chesapeake Bay anoxia. *Science, 254*:992–996.

volved clearing of old-growth forest for agriculture, which was substantially complete by 1920, at least in the watershed of Chesapeake Bay.[3]

Land-use changes are still occurring, most notably the intense "suburbanization" of much of the eastern seaboard, which was a hallmark of the twentieth century. At the same time, a few areas are being reforested, especially in New England, as farms are abandoned.

Among the other things we owe to the twentieth century is the idea of a "smooth, green carpet as a necessary adjunct to the perfect home" that is, a lawn. Thus, accompanying this profound shift in land usage has been the conversion of large areas of the eastern states to turfgrass. Moreover, the turfgrass component has been growing with the increase in suburban homes, golf courses, roads, and commercial development. In general, the rate of increase in turf plantings advances in lockstep with the increase in single-family home construction, as shown in the following examples:

- By the mid 1990s, the World Resources Institute[4] reported that 10% of the state of Maryland was composed of turfgrass, in the form of lawns, road medians and shoulders, golf courses, and other ornamental plots.
- Lawns make up over 2 million acres (640 acres = 1 mi^2) of the entire state of Pennsylvania, according to research data from Penn State University.[5]
- Overall, there were more than 25 million acres of lawns in the United States as of 1999.[6]

Question 1:   Refer to the conversion tables inside the front cover and convert the lawn area in the United States to (a) square miles, (b) square meters, and (c) hectares.

TURF'S COMPOSITION AND IMPACT

Turf is grass, but what is grass and what is its environmental impact? Turf commonly comes from one of two sources: (1) grass seed sown on a plot of dirt, or (2) pre-grown turf rolls (also called sod) produced on turf "farms." Turf farms (Figure 12-1) are lucrative sources of revenue: Consider these examples from the Pacific Northwest.

Washington State
Turfgrass Sod, Sprigs, or Plugs:
Revenue in 1988—-$6,608,000
Revenue in 1998—-$9,900,000

Willamette Valley, Oregon
Grass Seed:
1988   332,610 acres   Sales: $190 million
1997   410,510 acres   Sales: $300 million

---

[3]See U.S. Energy Information Administration. 1996. *Emissions of Greenhouse Gases in the United States,* U.S. Dept. of Energy, Washington, DC, p. 65, Figure 11. This publication includes maps showing the approximate extent of forest cover in the United States for 1620, 1850, 1920, and 1992. The eastern United States was virtually cleared of virgin forest between 1850 and 1920 and has been partially reforested since 1920. Even so, less than half the forest cover of 1620 remains.
[4]World Resources Institute. 1997. *World Resources 1996–97: A Guide to the Global Environment* (New York: Oxford University Press).
[5]Penn State University (data on turf). http://www.cas.psu.edu/docs/cashome/progper/turfgrass.html.
[6]Ibid.

FIGURE 12-1 Growing sod in the desert in Las Vegas, NV. (©Robert Holmes/CORBIS)

Most new housing uses turf rolls from sod farms to give the house or commercial building an "instant lawn." Turf rolls are simply rolled onto bare dirt and lightly compacted. The environmental impact of lawns and other turf plantings is complex and can be either positive or negative, depending on what the turf replaces, how the turf is maintained, and one's assumptions.

According to the Turfgrass Producers International (TPI),[7] an industry trade group, "sod cools and cleans the atmosphere by reflecting the sun's heat and absorbing noises, carbon dioxide, and harmful pollutants." Moreover, "It releases valuable oxygen and moisture into the air we breath [sic]." Lawns grown from sod rather than from grass seed also are asserted to improve sediment control and require less watering and pesticides.

While Americans use more than 50 million pounds of pesticides, fertilizer, and herbicides on their lawns annually, Penn State University research shows that around 65% of runoff samples from scientifically managed turfgrass test plots contain no pesticides or fertilizers, with the remaining samples yielding trace amounts that fall within present drinking-water standards. Thus, their results suggest that, managed properly, turf can play a role in minimizing soil erosion and water pollution.

According to the Professional Lawn Care Association of America (PLCAA),[8] lawns do the following:

- Produce oxygen as the plants photosynthesize; "625 square feet of lawn provides enough oxygen for one person for an entire day"
- Cool the temperature; "On a block of eight average houses, front lawns have the cooling effect of 70 tons of air conditioning"
- Control allergies by controlling dust and replacing plants to which many people are allergic
- Absorb gaseous pollutants CO_2 and SO_2
- Trap particles (up to 12 million tonnes annually)
- Protect homes from fires
- Protect water quality by filtering runoff

---

[7]Turf Producers International. Address: 1855-A Hicks Rd., Rolling Meadows, IL 60008 (http://www.turfgrasssod.org/).
[8]Professional Lawn Care Association of America (PLCAA) (http://www.plcaa.org/).

Question 2:   Evaluate the first claim, using critical thinking principles. Certainly turf produces more oxygen than dirt or asphalt, but how does turf compare with forest or other land uses? What additional information do you need to assess this claim?

Question 3:   Does the statement on cooling tell you what the cooling effect of turf is being compared to? Asphalt? Forest? Is the statement accurate? Precise? Clear? Ambiguous? What additional information do you need? Discuss.

Question 4:   Grass allergies are a widespread allergy. Is this mentioned in the third claim? (See *For Further Study* questions below.)

We will examine some of the other claims later in this discussion. Now we will consider the following points and determine whether there is a problem associated with expanding turf area.

1. Even though there are over 5000 species of grasses worldwide, sodded lawns contain very few species, and most warm-season grasses are dominated by a single species.
2. Although sod is clearly better at erosion control than bare dirt, much turfgrass has replaced forest, which is much more efficient at all the attributes described by the TPI. Sod has replaced pasture and other agricultural land as well (see *For Further Study* section for details).
3. While Penn State agricultural scientists were able to grow turf under controlled conditions with little toxic runoff, most private lawns are over-fertilized and receive inappropriate doses of insecticide and herbicide. For example, an annual nitrogen application of 3 to 4 lbs/1000 ft^2 will likely be necessary for new turf estab-

lished on a site devoid of organic matter and most plant nutrients, like most new suburban home sites. For the first few months following turf seeding or sodding, nitrate leaching can occur. High applications of nitrate-containing fertilizers made during late summer or early fall, if followed by heavy rain, can also promote nitrate leaching. Thus, improper fertilization of lawns can release substantial nitrate and phosphate into streams, fueling algal growth that leads to depletion of oxygen (hypoxia or anoxia). This may cause fish kills and contribute to infestations such as the *Pfiesteria piscicida* outbreak in 1997 that caused the closure of several streams tributary to Chesapeake Bay.[9]

4. Whether the lawn is from turf or seed, it should be watered when dry, mowed at regular intervals, and fertilized twice a year, according to the TPI.

5. Most importantly, increases in turf areas are almost always associated with sprawl development (see Issues 5 and 6), which is replacing agricultural land and woodland. Agricultural land is critically important as world and domestic populations grow (see Issue 1 and consult the website of the American Farmland Trust[10]). Moreover, woodland is irreplaceable as wildlife habitat, is essential for aesthetics, and can help counter global warming induced by greenhouse gases.

6. "Caring" for turfgrass has not-so-quietly become one of the nation's most polluting activities. Since most lawn-care devices (mowers, trimmers, blowers, weed-whackers, and the like) use two-cycle engines, they burn oil—lots of it. In fact, a gasoline-powered lawnmower run for an hour puts out about the same amount of smog-forming emissions as 40 new cars run for an hour, according to the State of California's Air Resources Board (ARB).

Turf poses more than environmental or land-use issues. It has growing economic implications as well. As of 1995, at least 500,000 workers were employed directly in turf maintenance and care,[11] and by 1998 there were 60,000 in California alone. That number is probably growing as an increasing number of suburban households with working spouses find themselves with a decreasing amount of time to take care of their lawn. In fact, U.S. homeowners are turning to lawn, landscape, and tree care professionals in record numbers, spending an all-time high $17.4 billion on outdoor "home improvement" in 1999. More than 26 million households hired a "green professional," a 23% increase over the previous year, and that number could have reached 29 million in 2000, according to a Gallup survey.

Lawn-care devices themselves, such as blowers, mowers, and trimmers ("weed whackers"), most of them gas-powered, represent a $5 billion per year growth industry. For example, Higgins[12] reported that in 1996, 6.8 million power mowers were sold, of which 1.45 million were riding/lawn tractors. In 1997, 7.03 million were sold, of which 1.5 million were riding/lawn tractors.[13] Is this cause for concern or congratulation?

---

[9]You may assess the relative environmental impact of urban/suburban, pasture, cropland, and forest land uses by accessing the EPA Chesapeake Bay Program Web site, available from www.epa.gov/r3chespk/ and www.chesapeakebay.net/bayprogram.

[10]www.farmland.org.

[11]Penn State University (data on turf) (http://www.cas.psu.edu/docs/cashome/progper/turfgrass.html).

[12]Higgins, A. Cutting edge. *Washington Post,* August 15, 1996.

[13]Wee, E.L. Homeowners feel a powerful tractor pull. *Washington Post,* May 21, 1997.

Question 5: According to the TPI,[14] a "well-maintained" 10,000 ft² lawn will generate around 1 ton (900 kg) of grass clippings annually. On average, what is the weight of grass clippings (in pounds) produced per square foot of lawn per year? Restate your answer as kilograms per square meter.

Question 6: Based on this value, how many kilograms of grass clippings are produced from Pennsylvania lawns each year?

In many states, much of this yard waste probably ends up in landfills, since according to EPA, grass clippings make up at least 10% of landfill waste in the United States. We will discuss this point later. Table 12-1 provides some additional statistics on turfgrass from the Bureau of the Census,[15] based on surveys taken every five years. The latest survey was to have been taken in 1997, but as of early 2001 was still unavailable.

TABLE 12-1 ■ U.S. Turfgrass Sod Production Industry[16]

Year	Acres in Production	Sales ($)
1974	85,164	97,159,000
1978	119,725	174,240,000
1982	124,588	210,510,000
1987	184,070	391,635,000
1992	218,161	471,640,000

---

[14]http://www.turfgrasssod.org/.
[15]U.S. Bureau of the Census. 1996. *Statistical Abstract of the U.S.: 1996* (116th ed.), Washington, DC: U.S. Government Printing Office.
[16]Ibid.

Question 7: What is the annual rate of increase of land used for sod cultivation from 1974 to 1982? (HINT: Use the formula $k = (1/t)\ln(N/No)$.)

Question 8: What is the annual rate of increase of land used for sod cultivation from 1982 to 1992?

Question 9: Based on the average annual rate of increase you calculated for 1982-92, project the acres in production in 2020. (HINT: Use $N = N_0 \times (e)^{kt}$, presented in *Using Math in Environmental Issues,* pages 14–15).

Question 10: A study by Beard and Green[17] maintains that turfgrass (*Cynodon,* or Bermuda grass) is effective at reducing ground temperature in summer (Table 12-2). Clearly, green turf results in lower temperatures than brown turf or synthetic turf. What special treatment is required to keep the turf green in the middle of the blistering summers of College Station, Texas?

---

[17]Beard, J.D., & R.L. Green. 1994. The role of turfgrasses in environmental protection and their benefits to humans. *Journal of Environmental Quality,* 23:452–460.

TABLE 12-2 ■ Temperature Comparisons of Four Types of Surfaces on August 20 in College Station, Texas[18]

Type of Surface	Maximum Daytime Surface Temperature (°C)	Minimum Nocturnal Surface Temperature (°C)
Green, growing *Cynodon* turf	31	24
Dry, bare soil	39	26
Brown, summer-dormant *Cynodon* turf	52	27
Dry, synthetic turf	70	29

Question 11: According to the PLCAA,[19] most lawns require 1 inch of rain or irrigation water per week. Calculate the amount of water necessary to irrigate 5000 ft² of lawn each week, assuming there is no rain. Express the amount in liters per day.

Question 12: How does this figure compare with the U.S. 1990 per capita water consumption of 1340 gallons per day[20]? (Convert your answer to liters).

ALLERGIES, INDIVIDUAL RIGHTS, AND LEAF BLOWERS

Many southwestern cities such as Albuquerque, New Mexico; Tucson, Arizona; Las Vegas, Nevada; and El Paso, Texas have reported increases in residents with allergies. In Tucson "residents have twice the national rate of respiratory allergies."[21] Experts attribute the increases to people who move to the west from eastern states and bring a taste for lawns and eastern trees with them.

Some cities are beginning to ban certain plant species. For example, Tucson has banned Bermuda grass (*Cynodon*), a key species in many lawns, and many western cities discourage turf plantings while encouraging "xeriscaping," that is, planting native drought-resistant plants in lawns to cut down on water demand. However, many residents object that their "rights" are being violated. When Albuquerque's City Council banned several types of trees, a resident responded, "I get pretty sick of people coming in from

---

[18]Ibid.

[19]http://www.plcaa.org/.

[20]U.S. Bureau of the Census. 1996. *op. cit.*

[21]Mendoza, M. Southwestern communities find greening desert is something to sneeze at. *Washington Post*, November 10, 1996.

the east saying that this is a desert and we have to live like it's a desert. We do not have to live like it's a desert. I love trees and I choose to live near them."[22]

Question 13: Do you think that city governments have an obligation, or the authority, to tell residents what they can or cannot plant? Do you think that anyone should have a "right" to plant whatever they like on their own property? Defend your reasoning.

Question 14: Healthcare costs consistently rise faster than the rate of inflation, and part of the cost of healthcare is borne by people with severe allergies. Do you feel that those who insist on their "right" to plant whatever they choose on their property should be required to pay for the added burden on the healthcare system that their actions cause or contribute to? How? Explain your reasons. Is this an environmental issue? Why or why not?

Question 15: Do you believe that governmental attempts to regulate ornamental vegetation like lawns and trees in order to protect allergy sufferers or water supplies represents an unnecessary regulatory burden by "Big Government" on U.S. citizens? State and defend your reasons.

RIDING LAWNMOWERS

As mentioned above, 1.5 million riding lawnmowers were sold in 1997, up 3% from 1996. Top-of-the-line riding mowers can cost $3000 to $4000 and most are financed at 10% down with up to 4 years to pay, rather than being purchased with cash. Moreover, they come with optional equipment, including stainless steel hubcaps for $130 and an FM-cassette stereo for $149.[23]

---

[22]Ibid.
[23]Wee, E.L. Homeowners feel a powerful tractor pull. *Washington Post*, May 21, 1997.

Question 16: Calculate how much money Americans spent on riding lawnmowers in 1997, assuming an average sale price of $3000. (Or check out the advertisements for riding lawnmowers in your local newspaper and derive your own average sales price.)

Question 17: Assume buyers finance the purchase with a credit card. Calculate how much interest alone these buyers will pay, assuming a loan for 90% of the purchase price of $3000 (or the price you derived), spread out over 4 years at an interest rate of 12%. (You can use the compound interest equation explained in *Using Math in Environmental Issues,* pages 14–15).

GAS-POWERED LEAF BLOWERS

In June 2000, the California ARB submitted a report to the California Legislature on leaf blowers.[24] Here is a summary of that report.

The gas-powered leaf blower was invented in the early 1970s by Japanese engineers and introduced to the United States as a lawn and garden maintenance tool. It gained little acceptance until drought conditions in California during the late 1970s led to prohibitions of water use for many garden clean-up tasks. These water-saving measures led to the proliferation of leaf blowers. In 1990, 800,000 were sold nationwide. In 1998, industry shipments of gasoline-powered handheld and backpack leaf blowers increased 30% over 1997 shipments to 1,868,160 units nationwide. In California there are more than 400,000 gasoline-powered leaf blowers, plus approximately 600,000 electric leaf blowers, that were operated an estimated 114,000 hours per day as of mid-2000.

Soon after the introduction of the leaf blower into the United States, they began to generate citizen complaints, mainly for noise, pollution, and dust. Beverly Hills and Carmel-by-the-Sea, two of the wealthiest and most exclusive municipalities in California, banned the devices during the 1970s. By 2000, 100 cities had banned or regulated leaf blowers.

Arizona and New Jersey have also considered regulating leaf blowers, and five other states have at least one city with a leaf blower ordinance. The Orange County, CA Grand Jury and the Palo Alto City Manager's Office both weighed in on the issue of leaf blowers, and proposed that agencies stop using gasoline-powered leaf blowers in their maintenance and clean-up operations. The major findings of these agencies are similar and presented below in Table 12-3.

The City of Palo Alto subsequently voted to ban the use of fuel-powered leaf blowers throughout the city as of July 1, 2001.

---

[24]http://www.arb.ca.gov/msprog/leafblow/leafblow.htm.

Table 12-3 ■ Major Findings of Orange County Grand Jury and City of Palo Alto

Orange County Grand Jury Report (1999)	City of Palo Alto City Manager's Report (1999a)
(1) Toxic exhaust fumes and emissions are created by gas-powered leaf blowers.	(1) Gasoline-powered leaf blowers produce fuel emissions that add to air pollution.
(2) The high-velocity air jets used in blowing leaves whip up dust and pollutants. The particulate matter (PM) swept into the air by blowing leaves is composed of dust, fecal matter, pesticides, fungi, chemicals, fertilizers, spores, and street dirt, which consists of lead and organic and elemental carbon.	(2) Leaf blowers (gasoline and electric) blow pollutants including dust, animal droppings, and pesticides into the air, adding to pollutant problems.
(3) Blower engines generate high noise levels. Gasoline-powered leaf blower noise is a danger to the health of the blower operator and an annoyance to the nonconsenting citizens in the area of usage.	(3) Leaf blowers (gasoline and electric) do produce noise levels that are offensive and bothersome to some individuals.

The ARB report explained the advantages and environmental disadvantages of leaf blowers.

Small, two-stroke gasoline engines have traditionally powered leaf blowers, and most still do today. The two-stroke engine has several attributes that are advantageous for applications such as leaf blowers. They are lightweight in comparison to the power they generate, and operate in any position. Multi-positional operation is made possible by mixing the lubricating oil with the fuel; the engine is, thus, properly lubricated when operated at a steep angle or even upside down.

A major disadvantage of two-stroke engines is high exhaust emissions. Typical two-stroke designs feed more of the fuel/oil mixture than is necessary into the combustion chamber. Through a process known as scavenging, the incoming fuel enters the combustion chamber as the exhaust is leaving. This timing overlap of intake and exhaust port opening can result in as much as 30% of the fuel/oil mixture being exhausted unburned. Thus, exhaust emissions consist of both unburned fuel and products of incomplete combustion. The major pollutants from a two-stroke engine are, therefore, oil-based particulates, a mixture of hydrocarbons, and carbon monoxide. A two-stroke engine forms relatively little oxides of nitrogen emission. . . .

Some of the hydrocarbons in fuel and combustion by-products are themselves toxic air contaminants, such as benzene, 1,3-butadiene, acetaldehyde, and formaldehyde. . . . Data do not exist to allow reliable estimation of toxic air contaminant emissions from small, two-stroke engine exhaust.

Question 18: Summarize the potential emissions from two-stroke engines and their possible ill effects. Did the ARB conclude that they could definitively assign a number value to the emissions from two-stroke lawn-care devices?

Below is Table 12-4, which summarizes the ARB's best estimate of pollution from Lawn Care Devices, and, based on regulations effective in 2000, the contributions of leaf blowers to the state's air pollution in 2010.

Table 12-4 ■ Inventory of Leaf Blower Exhaust Emissions (tons per day) in California

	Leaf Blowers 2000	Leaf Blowers 2010	All Lawn & Garden, 2000	All Small Off-Road, 2000
Hydrocarbons, Reactive	7.1	4.2	50.24	80.07
Carbon Monoxide (CO)	16.6	9.8	434.99	1046.19
Fine Particulate Matter (PM10)	0.2	0.02	1.05	3.17

Question 19:   Based on the table above, how much would the 2000 regulations, if properly implemented, reduce air pollution from leaf blowers?

Question 20:   Is it a proper role of government to monitor devices produced by corporations and to set standards for their operation, or should manufacturers be required to demonstrate that their products will cause no harm to the environment before they are allowed to produce and distribute these devices? Explain your answer.

DUST

Two other sources of emissions from leaf blowers are "fugitive"/resuspended dust, and noise. We will consider these next.

Leaf blowers are designed to move relatively large materials such as leaves and other debris, and hence will also entrain much smaller particles, especially those below 30 micrometers in diameter, which are not visible to the naked eye. The leaf blower moves debris by pushing large volumes of air at a high wind speed, typically 150 to 280 miles per hour (hurricane wind speed is >117 mph).

The Orange County Grand Jury alleged that substances such as fecal material, fertilizers, fungal spores, pesticides, herbicides, pollen, and other biological substances could be found in the dust resuspended by leaf blower usage. The ARB estimated that typical streets and sidewalks could contain about 3 grams of fine sediment per square meter, which would be blown into the air by a leaf blower. In addition, the chemical analysis of paved road dust showed small percentages of the toxic metals arsenic, chromium, lead, and mercury, and contributions from tire and brake wear particles. Fine latex particles from wearing of tires are a known human allergen, but the effect of resedimenting the material is unknown.

Question 21: The parking areas adjoining the Potomac Mills Mall mentioned in Issue 5 measure approximately 3,000,000 ft². Based on an average value of 3g/square meter of dust, how must dust could be resuspended by "cleaning" the parking lots using leaf blowers? Report your answer in grams and pounds. (First convert the feet to meters.)

NOISE

In addition to damaging hearing, noise may cause other adverse health impacts, including rest and sleep disturbance, changes in performance and behavior, annoyance, and other changes that may lead to poor health. In fact, leaf blower noise may have contributed to at least one murder. According to the Bergen (NJ) *Record,* in May 2000 "a woman killed her 74-year-old neighbor by repeatedly running him over with a car, after their latest dispute, which involved his use of a leaf blower, police said." The victim had complained to a neighbor earlier that the alleged assailant, wearing a dust mask, had threatened him earlier with a pitchfork due to the noise, emissions, and dust from the victim's leaf blower.

According to the ARB, long-duration, high-intensity sounds are the most damaging and usually perceived as the most annoying. High-frequency sounds, up to the limit of hearing, tend to be more annoying and potentially more hazardous than low-frequency sounds.

Figure 12-2 shows some commonly measured noise levels and some reference values from leaf blowers, from the ARB.

Perceived Sound Level	dB	µPa	Examples	Leaf Blower Reference
PAINFULLY LOUD	160	2×10^9	fireworks at 3 feet	
	150		jet at takeoff	
UNCOMFORTABLY LOUD	140	2×10^8	threshold of pain	OSHA limit for impulse noise
	130		power drill	
	120	2×10^7	thunder	
VERY LOUD	110		auto horn at 1 meter	90–100 dB leaf blower at operators ear
	100	2×10^6	snowmobile	
	90		diesel truck, food blender	90 dB OSHA permissible exposure limit
MODERATELY LOUD	80	2×10^5	garbage disposal	
	70		vaccum cleaner	82–75 dB Leaf blower at 50 feet
	60	2×10^4	ordinary conversation	
QUIET	50		average home	
	40	2×10^3	library	
VERY QUIET	30		quiet conversation	
	20	2×10^2	soft wisper	
BARELY AUDIBLE	10		rustling leaves	dB = decibels
	0	2×10^1	threshold of hearing	µPa = micro Pascals

FIGURE 12-2 Noise levels from common sources and some leaf blower comparisons. (California ARB)

Based on extrapolations of 20-year-old EPA data, it is possible that at least three million people nationwide are exposed to leaf blower noise at annoying levels. For California, the figure could be around 300,000, based on the urban/rural population ratio and scaling of population values, but since 60,000 persons were employed by the lawn-care industry in the state in 1998, at least that number is routinely exposed to excessive noise from leaf blowers.

Legislation

The Federal Noise Control Act of 1972 established a national policy "to promote an environment for all Americans free from noise that jeopardizes their public health and welfare." The EPA is charged with implementing this law.

About 13% of Californians live in cities that ban the use of leaf blowers, and six of the ten largest California cities have ordinances that restrict or ban leaf blowers. All together, to summarize, about 100 California cities have ordinances that restrict either leaf blowers specifically or all gardening equipment generally, including cities with bans on leaf blower use.

Summary

California ARB concluded that "lawn and landscape contractors, homeowners using a leaf blower, and those in the immediate vicinity of a leaf blower during and shortly after operation, are exposed to potentially high exhaust, fugitive dust, and noise emissions from leaf blowers on a routine basis."

Question 22: Use the information in Table 12-5 below and compare the impact of leaf blowers to the pollution produced from vehicles (NOTE: "Older light-duty vehicles" are extremely rare today and are included for comparative purposes).

Table 12-5 ■ Air Pollution from Leaf Blowers Compared to Motor Vehicles (ARB report)

	Exhaust Emissions, g/hr	Exhaust Emissions New Light Duty Vehicle, * g/hr	Exhaust Emissions Older Light Duty Vehicle, ** g/hr
Hydrocarbons	199.26	0.39	201.9
Carbon Monoxide	423.53	15.97	1310
Particulate Matter	6.43	0.13	0.78
Fugitive Dust	48.6–1031	N/A	N/A

*New light duty vehicle represents vehicles one year old, 1999 or 2000 model year, driven for one hour at 30 mph.
**Older light duty vehicle represents vehicles 1975 model year and older, precatalytic vehicle, driven for one hour at 30 mph.

Question 23: Approximately 17 million new motor vehicles are sold in the United States each year, each of which contains a catalytic converter that costs approximately $300. Calculate the total amount Americans pay for catalytic converters to take pollution out of the air and then comment on whether it makes economic sense to put that pollution back in the air in the form of toxics from leaf blowers. Defend your answer.

ARB concludes that leaf blower operators could be exposed to serious levels of particulates and other forms of toxics over the course of months to years. Here is their finding: "While leaf blower operators would not be expected to spend significant amounts of time within . . . a particulate cloud, the day-in-day-out exposure to [significant levels of] PM10 could result in serious, chronic health consequences in the long term."

Question 24: Should the healthcare system in this country have to bear the burden of the air pollution from leaf blowers' effects on operators and bystanders? Explain your answer.

Question 25: Read the following summary from ARB and state to what degree you feel whether these devices should be regulated, or whether manufacturers should be responsible for the health effects of operating these devices.

> The evidence seems clear that quieter leaf blowers would reduce worker exposures and protect hearing, and reduce negative impacts on bystanders. Costs and benefits of cleaning methods have not been adequately quantified.
>
> Fugitive dust emissions are problematic. The leaf blower is designed to move relatively large materials, which requires enough force to also blow up dust particles. Banning or restricting the use of leaf blowers would reduce fugitive dust emissions, but there are no data on fugitive dust emissions from alternatives, such as vacuums, brooms, and rakes. We need a more complete analysis of potential health impacts, costs and benefits of leaf blower use, and potential health impacts of alternatives.
>
> Some have suggested that part of the problem lies in how leaf blower operators use the tool, that leaf blower operators need to show more courtesy to passersby, shutting off the blower when people are walking by. Often, operators blow dust and debris into the streets, leaving the dust to be re-suspended by passing vehicles. Interested stakeholders, including

those opposed to leaf blower use, could join together to propose methods for leaf blower use that reduce noise and dust generation, and develop and promote codes of conduct by workers who operate leaf blowers. Those who use leaf blowers professionally would then need to be trained in methods of use that reduce pollution and potential health impacts both for others and for themselves.

FOR FURTHER STUDY

■ Access the Web page of the EPA Chesapeake Bay Program[25] and compare the nutrient and sediment pollution from forest, urban/suburban areas, pasture, and cropland. Discuss the implications of the spread of urban/suburban areas. Assess the role of turf plantings.

FIGURE 12-3 The Reserve Gold Course in Pawley's Island, SC. This course was designed to be environmentally sensitive and has much less turf than a typical golf course. (Courtesy of The Reserve Golf Course)

---

[25]www.epa.gov/r3chespk/ and www.chesapeakebay.net/bayprogram.

- Research the impact of the proliferation of riding lawnmowers on the healthcare system, in light of healthcare experts' estimates that over 25% of adult Americans are obese.
- Contact the city government of Albuquerque, Tucson, or one of the other cities mentioned above and find out about their regulations on turf plantings. Ask your physician for her or his opinion about allergies caused by turf or trees.
- Contact your city or county department of waste management and find out whether they accept lawn debris or waste.
- At the EPA's website,[26] find out how serious emissions from gasoline-powered lawn-care devices are and what regulations are in effect or proposed to control emissions from gasoline-powered lawn-care devices. In 1997, the County of Los Angeles restricted the use of gasoline-powered lawn-care devices and banned them in 1998. Research this issue from articles in the Los Angeles *Times* or the websites for Los Angeles County and describe the reasons the local officials took this step. What was the response of the lawn-care industry?
- Contrast the impact of gasoline-powered with electric lawn-care devices. What advantages do electric devices provide? What disadvantages?
- Golf courses typically replace fields or forests with turf. Some newer courses (Figure 12-3), are reducing the area of turf. Research this issue. Check the golf courses in your area. Are they constructed and maintained in a sustainable way?

---

[26]U.S. Environmental Protection Agency (http://www.epa.gov).

THE IMPACT OF INTERJURISDICTIONAL WASTE DISPOSAL: TRUCKIN' TRASH

KEY QUESTIONS

- How much municipal solid waste (MSW) is transported across political boundaries for disposal?
- What is the practice's impact on roads (Figure 13-1), bridges, and traffic safety?
- How much does this practice rely on cheap fuel, the U.S. Constitution, and the interstate highway system?

BACKGROUND

An April 2000 press release from the New York City Mayor's Office read as follows:

"New York City is committed to closing the Fresh Kills landfill (Figure 13-2) by December 31, 2001," Mayor Rudy Giuliani said. "This plan brings New York into the 21st century with a waste management system that ends the City's 50 years of reliance on the Fresh Kills Landfill." To handle the new plan, the City will develop five waste export facilities which will allow the Department of Sanitation to shift approximately 13,000 tons of solid waste now disposed at Fresh Kills Landfill on a daily basis, to out-of-City disposal sites.

FIGURE 13-1 Truck hauling trash along a rural road. (D. Abel)

FIGURE 13-2 Garbage at New York's Fresh Kills landfill, one of the world's largest landfills. (Owen Franken/CORBIS)

An earlier (1997) release announced:

> The New York City Department of Sanitation has awarded a three-year contract to begin exporting as many as 1,750 tons of solid waste a day—or over 530,000 tons each year—from the Bronx. Instead of coming to Fresh Kills, this waste will be transported to a landfill in Waverly, Virginia. It will cost the city $51.72 per ton of waste, less than the experts had anticipated.

MANAGEMENT OF MSW

In the last fifteen years, the United States has seen a major shift in the management of municipal solid waste (MSW; Table 13-1).

Two-thirds of the nation's landfills have closed as regulations governing disposal of MSW radically tightened. In the early 1980s there were approximately 13,000 municipal landfills in the country. The number of landfills dropped to 8000 in 1988 and to 2400 by 1996. The capacity, however, has remained relatively constant.

The cost of waste management rose significantly during the 1980s,[1] but during the 1990s, in most areas of the country the cost leveled off or even declined, mainly due to the construction of giant regional landfills that charge relatively low fees for disposal. So much additional capacity has been added that there is now an excess of disposal capacity in many areas.[2] From 1986 to 1996 construction of incinerators mushroomed and recycling grew as well. According to the Congressional Research Service,[3] by 1995 incinerators were burning 16% of the nation's MSW. But over the past five years incinerators have become less favored due to the high construction and operating costs involved, as well as environmental concerns about air quality. Moreover, incinerators leave a residue of ash, sometimes containing toxic substances, that must be buried somewhere.

In the first half of the 1990s, recycling and composting were the fastest growing methods of waste handling, accounting for 27% of waste management in 1995, up from

[1]Cunningham, W.P. & W.B. Saigo. 1997. *Environmental Science: A Global Concern* (Dubuque, IA: McGraw-Hill/Wm.C Brown Publishers).
[2]McCarthy, J.E. 1995. *Interstate Shipment of Municipal Solid Waste: 1998 Update.* Congressional Research Service report 95-570 ENS. Washington DC. Access available from www.cnie.org/.
[3]Ibid.

TABLE 13-1 ■ Composition of MSW in Percent by Weight[4]

	%
Paper and paperboard	37.5
Yard waste	17.9
Glass	6.7
Metals	7.7
Iron	6.3
Aluminum	1.4
Plastics	8.3
Wood	6.3
Food	6.7
Other (textiles, rubber, leather, etc.)	8.9

10% in 1986. Nearly 9000 local governments have begun curbside collection of recyclable materials and 3300 have composting programs for yard waste.

Question 1: In 1997, MSW totaled approximately 209 million tons.[5] Calculate the per capita municipal waste in 1997. (1997 population = 268 million.)

According to Supreme Court decisions based on the Constitution's Commerce Clause, Congress has sole authority to regulate interstate commerce (see below). Since the late 1980s, Congress has considered, but never enacted, bills that would allow states to impose restrictions on interstate waste shipments. Without congressional authorization, states do not have the right to ban waste shipments.

INTERJURISDICTIONAL TRUCKING OF SOLID WASTE

With the advent of giant regional landfills, the U.S. interstate highway system has become indispensable for moving the nation's consumer goods and each year moves a greater share of its MSW. The situation has been helped along by the low transportation fuel costs in the United States.

Today, as a result of low fuel prices, an integrated rail network, and the interstate highway system, it is common for lettuce from California to be served in salads in New York, and for tomatoes from Mexico and Florida to be distributed around the eastern and midwestern states. But it is also becoming common for municipal waste from New York City to be transported over 1000 km for disposal in huge commercial landfills in Pennsylvania, Virginia, and other states; for industrial waste from Michigan to be hauled to the Dakotas for dumping; and for waste from Los Angeles to be transported by truck and rail into the Mojave Desert (Table 13-2).

---

[4]Blatt, H. 1997. *Our Geologic Environment.* (Upper Saddle River, NJ: Prentice Hall).
[5]Macauley, M.K., S.W. Salant, M.A. Walls, & D. Edelstein. 1993. *Managing Municipal Solid Waste: Advantages of the Discriminating Monopolist.* Resources for the Future Discussion Paper Abstract ENR93-05 (Washington, DC: Resources for the Future).

TABLE 13-2 ■ MSW Exporters and Importers[6] (1993)

Net Exporters	Millions of Tons	Net Importers	Millions of Tons
New York	3.7	Pennsylvania	4.8
New Jersey	1.6	Virginia	1.5
Missouri	0.9	Ohio	1.3
Washington	0.7	Oregon	0.8
Washington, D.C.	0.6	Connecticut	0.8
Rhode Island	0.6	Indiana	0.7
Iowa	0.3	Kansas	0.7
Ontario	0.25	New Hampshire	0.5
Texas	0.23	West Virginia	0.4
Minnesota	0.2	Wisconsin	0.3

For example, since 1988 Pennsylvania has increased landfill capacity such that the state now has over 10 years worth of capacity at current disposal rates, much more if imported waste were excluded. However, even though the amount of Pennsylvania's waste disposed in landfills has decreased by nearly 2 million tons a year due to recycling, the amount of municipal waste coming into Pennsylvania from other states has increased 94% between 1988 and 1995.[7] Of 209 million tons of MSW generated annually, at least 18 million tons, or about 9%, was transported across state boundaries for dumping in 1997 (Figure 13-3).

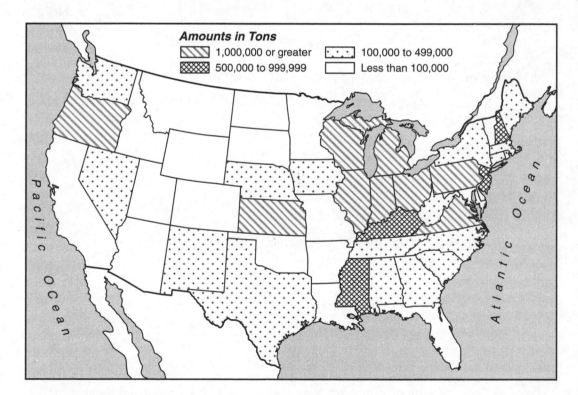

FIGURE 13-3 Interstate waste imports, 1997 (tons). (Congressional Research Service[8])

[6]McCarthy, J.E. 1995. *Interstate Shipment of Municipal Solid Waste: 1995 Update.* Congressional Research Service report 95-570 ENS, Washington, DC.

[7]Zero Waste America (http://www.zerowasteamerica.com).

[8]McCarthy, James E. 1998. Congressional Research Service Report for Congress Interstate Shipment of Municipal Solid Waste: 1998 Update http://www/cnie/org/nle/waste-7.html.

Although this tonnage seems minuscule, the transportation costs alone (not including disposal) amount to over $500 million annually.[9]

Question 2:   Calculate the average transport cost per ton that municipalities were willing to pay in 1997 to ship worthless garbage across state lines, rather than deal with the problem in their own locality.

Question 3:   Speculate to what extent low fuel prices contribute to this practice.

Question 4:   Refer back to Issue 3 (carrying capacity) and evaluate to what extent exportation of MSW is an indicator of a region's having exceeded its carrying capacity.

Even as some localities attempt to block the importation of solid waste, other jurisdictions apparently welcome it. In Virginia's Charles City County, southeast of Richmond, a landfill with a capacity of 5000 tons per day provides the following benefits to the county: 51 new jobs, three reductions in property taxes, money for schools and enhanced police protection, and a guaranteed $1 million per year minimum "host" payment to the county. When the landfill is finally closed, the county will receive title to the land. The landfill has much more stringent groundwater protection than required by the state, including two liners to monitor and trap seepage ("leachate") from the landfill and thus prevent the contamination of groundwater or streams.[10] Much of the waste dumped in this landfill comes

---

[9]Resources for the Future (http://www.rff.org).
[10]Miles, F.H. 1995. Testimony before the U.S. House of Representatives Subcommittee on Commerce, Trade and Hazardous Materials, March 25, 1995. Federal Document Clearing House, Inc., Washington, DC.

from New Jersey and New York and is carried by trucks with capacities of 30 to 40 tons per load.

Question 5: Virtually all these trucks use Interstate 95. Assume each truck carries 35 tons of MSW and that the landfill accepts 5000 tons per day. How many truck trips on Interstate 95 to and from the landfill are required daily? (Remember that the trucks must return empty for a new load!) Speculate on the impact these shipments will have on highway congestion, wasted fuel, and traffic safety.

Trucking trash is slated to increase during the next decade, especially for those states on the eastern seaboard. In 2001 as you have seen, one of the nation's largest dumpsites, Staten Island's Fresh Kills landfill, closed, and 13,000 tons of garbage a day, almost 5 million tons a year, will need a new home. Montana's Senator Max Baucus[11] described it this way in a speech on the Senate floor in 1997:

> It means about 1200 trucks of garbage a day coming out of New York City, every one of them packed to the brim. Or, in other words, a convoy of trash trucks 12 miles long, 365 days a year. Where is it going to go? One thing is clear. New York will have virtually no way to get rid of its trash when Fresh Kills does close in the year 2001. The entire State of New York can take only about 1200 tons of New York City's trash each day and that means the rest, over 4 million tons a year, must go out of State.

Question 6: Consider the items in Table 13-1. List and discuss alternatives to New York City's apparent decision to ship its waste out of state. Justify your answer.

SUPREME COURT RULINGS ON GARBAGE

In addition to more stringent water quality laws and cheap fuel, federal court rulings have had a profound impact on local waste management programs. In three decisions since 1992, the U.S. Supreme Court has ruled that shipments of garbage, even though "value-less" in the Court's words, constitute acts of "commerce" and thus are protected under the U.S. Constitution from state or local laws banning the importation of municipal garbage. The Commerce Clause, Article 1, section 8 (3) states simply, "The Congress shall have the power to regulate commerce with foreign nations and among the several States." As a

---

[11]U.S. Senate speech by Max Baucus, available from http://www.askwitteachik.org/federal/s443cc.htm or from http://rs9.loc.gov/cgi-.

result, state and local governments may not prohibit private landfills from accepting waste from out of state, nor may they impose fees that discriminate on the basis of origin on waste disposal from another state.

In a landmark 1978 case, *City of Philadelphia v. New Jersey,* the U.S. Supreme Court ruled 7 to 2 that New Jersey could not bar MSW from Philadelphia (437 U.S. 617):

> New Jersey sought to impose upon out-of-state commercial interests the full burden of conserving the state's remaining landfill space. . . . The New Jersey statute cannot be likened to a quarantine law which bars articles of commerce because of their innate harmfulness, and not because of their origin. New Jersey concedes that out-of-state waste is no different from domestic waste but it has banned the former. . . . Today cities in Pennsylvania and New York find it expedient or necessary to send their waste into New Jersey for disposal . . . tomorrow New Jersey may find it expedient or necessary to send their waste into Pennsylvania or New York for disposal and then these states might claim the right to close their borders.

TRUCKING TRASH AND HIGHWAY SAFETY

More and more localities are using "transfer stations" to collect trash at a central location for shipment to other sites for final disposal, incineration, or recycling. This practice has been the result of closure of over two-thirds of the nation's solid waste landfills in recent years, as mentioned above.

Although trash hauling has been a bonanza for some railroads, most trash is hauled by truck. To get an appreciation of the impact of truck transfer stations, consider the Jack C. Lausch Transfer Station, which serves southeastern Pennsylvania's Lancaster County (county area: 946 mi^2 = 1525 km^2). Each day, more than 200 trash-bearing trucks arrive at the station, and over 800 tons of refuse are trucked away to other sites for final processing.[12]

Citizens across the country are becoming increasingly concerned about the safety of long-distance waste transport as well as the numbers of trash trucks. In Pennsylvania, officials randomly inspect trash trucks and have documented a history of safety violations, including the following:

■ In September 1997, 117 of 487 waste haulers inspected throughout the state were cited with one or more violations. And in November 1996, 192 of 849 trash trucks were cited with one or more violations.[13]

■ A number of fatal accidents involving trash haulers has begun to focus attention on the practice in the mid-Atlantic states. For example, in March 1997 the driver of a tractor-trailer carrying 77,000 lbs of MSW from New York City to a landfill in Virginia was charged with reckless driving in a crash that killed two people, injured seven, and demolished three vehicles. Police cited the driver with reckless driving on a rain-soaked section of Interstate 95 in Maryland. They also charged the driver with falsifying his logbook.[14]

---

[12]Plastics Resource website (http://www.plasticsresource.com/groups/consumers/sub_index.asp?of=tours).
[13]Pennsylvania Department of Environmental Protection. Nov. 1 1996 update, available from http://www.dep.state.pa.us/dep/deputate.
[14]Reid, A. Trucker charged in pileup. *Washington Post,* March 18 1997.

Question 7: To what extent are the dead and injured in this crash part of the external cost of waste disposal (i.e., that part of the cost not reflected in the price paid to landfill operators, etc.)? In your answer discuss whether you think this is a "fair" question and state your reasons.

Question 8: Should tragedies like this (often called "accidents") be an integral part of any discussion on waste management? To what extent should people in jurisdictions that handle their waste problem by exporting it be held accountable for such disasters?

Question 9: When assessing the economics of waste reduction strategies, should monetary losses, deaths, and injuries avoided by reducing waste be considered? How? Explain your decision.

Finally, trucking trash is contributing to conflicts as localities and states try to raise the funds needed to maintain, repair, and rebuild roads and bridges. The issue of waste transport over public roads is clearly intermingled with air and water pollution issues, transportation and highway safety issues, as well as tax policy. As you can guess, we have barely scratched the surface of the issue of waste and waste "management."

FOCUS: THE WOODROW WILSON BRIDGE

The Woodrow Wilson Memorial Bridge, in MD, DC, and VA (Figure 13-4), connects Maryland and Virginia along Interstate 95 and is owned by the federal government. It is one of the most troublesome traffic "bottlenecks" on the entire length of Interstate 95, which is itself one of the nation's most heavily used interstate highways, connecting New

FIGURE 13-4 The Woodrow Wilson Bridge. (© 2002 Nate Bacon—Photographer)

England and New York with Florida. Traffic volumes across the Wilson Bridge are sobering: More than 220,000 passenger cars and 11,000 trucks cross it daily. A new bridge is being considered to replace the existing one before the present bridge deteriorates to the point that trucks would be banned, beginning in 2004. The price tag for the new bridge ranges between $1.3 and $3 billion. Trash trucks from New York and New Jersey routinely use the bridge to access the Charles City, Virginia landfill and others.

Question 10: Should trash trucks be banned, or taxed to help pay for bridge renovations? As noted above, Virginia imports more than 1.5 million tons of MSW annually. Virtually all of it crosses the Wilson bridge. How many trucks per day cross the bridge at present, assuming the average truck has a 35-ton load?

Question 11: After the Fresh Kills landfill closes, how many **additional trash trucks could cross the Wilson Bridge**, assuming they all head south? Should the companies that own the trucks be asked to contribute to bridge maintenance? Should New York City? How? Explain your reasoning.

FOR FURTHER STUDY

- Ohio is another state that bears a heavy impact from imported waste. Access the Ohio Environmental Protection Agency's Web and summarize the issues as seen by the Ohio EPA.
- Chattanooga, Tennessee is facing a MSW problem similar to many localities in that the cost to develop and maintain a landfill exceeds their ability (or desire) to pay. Their solution is to join with other surrounding localities and build a "regional" waste disposal facility. Research the issue[15] and evaluate the impact of such a regional solution on the area's road network.
- Determine how your own area handles solid waste. To what extent is recycling a solution to MSW? Determine how localities handle materials collected for recycling. Are they transported by road? Cite examples.

---

[15]Denton, L. Landfills are a mountain of worry. *Chattanooga Times,* April 15, 1997. Available from http://www.chattimes.com.

OIL RESOURCES AND RESERVES

KEY QUESTIONS

- How much oil is left?
- How fast are we using it up?
- How much oil is there in the Arctic National Wildlife Refuge (ANWR)?
- What are the issues surrounding drilling for oil in the ANWR?

BACKGROUND

Oil is the energy basis of modern industrial society, but for over fifty years after the first successful well was drilled in the mid-nineteenth century, oil use was insignificant. Henry Ford changed all that with his mass-produced gasoline-powered automobiles, combined with the invention of the electric starter in 1911. World War I generated immense demand for gasoline-powered vehicles. During the period of 1919 to 1949, oil gradually overtook coal as the most important fuel source in the United States (and accompanying the increase in oil use, the number of miles of road increased as well). Today, oil provides half of the U.S. energy requirement and nearly all of our transportation fuel.

By 2001, oil (or petroleum) use was approaching 80 million barrels (1 barrel = 42 U.S. gallons) per day globally, with the United States responsible for a quarter of that demand (nearly 20 million barrels per day).[1] Due to the precipitous decline in oil prices after 1982, oil demand, and fuel prices, in the United States by 1997 reached levels not seen since the late 1970s. In the past several years, largely because of renewed demand from depressed Asian economies coupled with the changing nature of the U.S. motor vehicle fleet, oil prices have risen sharply. The U.S. Energy Information Administration projects that U.S. and global oil demand will continue to increase through 2010.

ORIGIN, EXTRACTION, AND DISTRIBUTION OF OIL

Oil is not a renewable resource. The oil fields of today originated tens of millions of years ago when organic remains were buried within sediments in the absence of oxygen. The organic matter was subjected to a critical combination of pressure (caused by deep burial) and increased temperatures. Over millions of years, the organic matter reorganized into more volatile organic molecules that we call crude oil or petroleum, usually accompanied

---

[1]U.S. Energy Information Administration (http://www.eia.doe.gov/index.html).

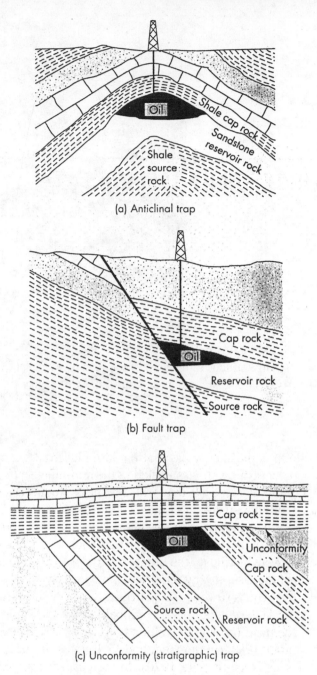

(a) Anticlinal trap

(b) Fault trap

(c) Unconformity (stratigraphic) trap

FIGURE 14-1 Typical oil and gas "traps." Oil, being lighter than water, floats to the top of the reservoirs. The oil is usually associated with natural gas as well. (Keller, E.A. 2000. *Environmental Geology* 8th ed., Fig. 15.10, page 411. Prentice Hall, Upper Saddle River NJ.)

by methane, which we call natural gas. To be extracted and used, the oil must migrate upward, out of its deeply buried *source bed,* and into a rock or a favorable geological structure that will trap the oil and prevent its escape (see Figure 14-1).

Sometimes pressure forces the oil all the way to the surface, where it forms *seeps,* which were known to Native Americans. The nineteenth century's oil discoveries came when "wildcatters" simply drilled holes into the rocks underlying oil seeps.

Once extracted from underground reservoirs, oil is processed to remove other fluids, such as saltwater and natural gas, and then it is sent to refineries. There the oil is heated in the absence of oxygen to break or "crack" the molecules into lighter forms, which emerge

FIGURE 14-2 Oil refinery at Martinez, CA, on an arm of San Francisco Bay. (©Ed Young/CORBIS)

as products such as gasoline, diesel fuel, or asphalt. Much, but not all, associated sulfur is usually removed at refineries as well. (The sulfur that is left in motor fuels when burned forms SOx, a toxic air contaminant, which cannot be removed by the present generation of catalytic converters.) And refineries are among the most polluting industrial sites in North America (Figure 14-2).

The oil may then be shipped by pipeline or oil tanker to destinations around the world.

HOW MUCH OIL IS LEFT?

Calculating the amount of oil present in all known world oil fields that can be extracted at a profit using present technology yields a value called *world proven oil reserves* (Table 14-1).

Agencies such as the American Petroleum Institute or the U.S. Department of Energy collect and publish such information. World proven oil reserves are less than the actual oil available, since technology may enhance oil recovery, the price may go up, and companies frequently underestimate the amount of oil in a field or may not publicize the actual number for competitive reasons. For example, in 1970 British Petroleum (BP) estimated its Forties North Sea oil field contained 1.8 billion barrels of proven reserves. However, by 1995, BP had produced 3.6 billion barrels and estimated that 2.8 billion barrels remained![2]

---

[2]The new economics of oil. *Business Week,* Nov. 3, 1997, pp. 140–148.

TABLE 14-1 ■ Proven Oil Reserves, 2000[3]

	At End 1979 Thousand million barrels	At End 1989 Thousand million barrels	At End 1999 Thousand million barrels	At End 1999 Share of total	At End 1999 R/P ratio
USA	33.7	33.6	**28.6**	2.8%	10.0
Canada	8.1	8.4	**6.8**	0.7%	9.3
Mexico	31.3	56.4	**28.4**	2.7%	24.5
Total North America	73.0	98.4	**63.7**	6.2%	13.8
Argentina	2.4	2.3	**2.7**	0.3%	9.1
Brazil	1.2	2.8	**7.3**	0.7%	18.1
Colombia	0.7	2.1	**2.6**	0.2%	8.5
Ecuador	1.1	1.5	**2.1**	0.2%	15.3
Peru	0.7	0.4	**0.4**	†	8.9
Trinidad & Tobago	0.7	0.5	**0.6**	0.1%	12.9
Venezuela	17.9	58.5	**72.6**	7.0%	65.2
Other S. & Cent. America	0.6	0.6	**1.2**	0.1%	25.0
Total S. & Cent. America	25.2	68.7	**89.5**	8.6%	37.7
Denmark	0.4	0.8	**1.1**	0.1%	9.7
Italy	0.6	0.7	**0.6**	0.1%	15.8
Norway	5.8	11.5	**10.8**	1.0%	9.3
Romania	n/a	n/a	**1.4**	0.1%	30.4
United Kingdom	15.4	4.3	**5.2**	0.5%	5.0
Other Europe	4.4	3.2	**1.6**	0.2%	13.3
Total Europe	26.6	20.5	**20.6**	2.0%	8.3
Azerbaijan	n/a	n/a	**7.0**	0.7%	69.5
Kazakhstan	n/a	n/a	**8.0**	0.8%	36.5
Russian Federation	n/a	n/a	**48.6**	4.7%	21.8
Turkmenistan	n/a	n/a	**0.5**	†	10.2
Uzbekistan	n/a	n/a	**0.6**	†	10.0
Other Former Soviet Union	n/a	n/a	**0.7**	0.1%	15.8
Total Former Soviet Union	67.0	58.4	**65.4**	6.3%	24.2
Iran	58.0	92.9	**89.7**	8.7%	69.9
Iraq	31.0	100.0	**112.5**	10.9%	*
Kuwait	68.5	97.1	**96.5**	9.3%	*
Oman	2.4	4.3	**5.3**	0.5%	15.9
Qatar	3.8	4.5	**3.7**	0.4%	14.7
Saudi Arabia	166.5	257.6	**263.5**	25.5%	87.5
Syria	2.0	1.7	**2.5**	0.3%	12.3
United Arab Emirates	29.4	98.1	**97.8**	9.4%	*
Yemen	–	4.0	**4.0**	0.4%	27.9
Other Middle East	0.2	0.1	**0.1**	†	9.1
Total Middle East	361.8	660.3	**675.7**	65.4%	87.0
Algeria	8.4	9.2	**9.2**	0.9%	20.6
Angola	1.2	2.0	**5.4**	0.5%	19.0
Cameroon	0.1	0.4	**0.4**	†	11.6
Rep. of Congo (Brazzaville)	0.4	0.8	**1.5**	0.1%	14.1

(continued)

[3]British Petroleum: (http://www.bp.com/worldenergy/oil/index.htm)

TABLE 14-1 ■ *Continued*

Egypt	3.1	4.5	**2.9**	0.3%	10.0
Equatorial Guinea	–	–	†	†	0.3
Gabon	0.5	0.7	**2.5**	0.2%	20.1
Libya	23.5	22.8	**29.5**	2.9%	57.4
Nigeria	17.4	16.0	**22.5**	2.2%	30.6
Tunisia	2.3	1.8	**0.3**	†	10.1
Other Africa	0.1	0.6	**0.6**	0.1%	8.1
Total Africa	57.1	58.8	**74.9**	7.2%	28.2
Australia	2.1	1.7	**2.9**	0.3%	15.0
Brunei	1.8	1.4	**1.4**	0.1%	20.8
China	20.0	24.0	**24.0**	2.3%	20.6
India	2.6	7.5	**4.8**	0.5%	17.8
Indonesia	9.6	8.2	**5.0**	0.5%	9.7
Malaysia	2.8	3.0	**3.9**	0.4%	14.0
Papua New Guinea	–	0.2	**0.3**	†	9.4
Thailand	–	0.2	**0.3**	†	8.6
Vietnam	–	–	**0.6**	0.1%	5.7
Other Asia Pacific	0.4	0.4	**0.8**	0.1%	12.9
Total Asia Pacific	39.4	46.6	**44.0**	4.3%	16.3
TOTAL WORLD	**650.1**	**1011.7**	**1033.8**	**100.0%**	**41.0**
Of which: OECD#	98.6	119.1	**85.6**	8.3%	11.8
OPEC	434.0	764.9	**802.5**	77.6%	77.4
Non-OPEC‡	149.1	188.4	**165.9**	16.0%	13.6

*Over 100 years.
†Less than 0.05.
#1979 & 1989 exclude Central European members.
‡Excludes Former Soviet Union.
n/a not available.

If prices rise, marginal fields may become profitable and may be added to reserves. Chevron Oil Corporation estimated in 1991 that 6700 billion barrels of oil may be ultimately recoverable, assuming *a price of $60 per barrel in 1990 dollars.*[4]

Question 1:   According to the Department of Labor, the Consumer Price Index stood at 127.4 in January 1990, and reached 168.8 by January 2000. Therefore, by what percent had the CPI increased over that ten-year interval?

[4]Holyoak, A.R. Written communication.

Question 2:   To determine the 2001 price that would be equivalent to a 1990 price of $60 per barrel, multiply the 1990 minimum price by 1 + the percentage you just calculated, changed to a fraction. What would be the 2000 price needed to "guarantee" 6700 million barrels of ultimately recoverable oil?

Question 3:   How many of the planet's citizens would be able to afford oil at this price, in your view? Explain your answer and cite evidence.

WHY HAVE OIL RESERVES GROWN?

Over the past several decades, technology has significantly enhanced discovery of oil and gas fields as well as ultimate recovery of the oil and gas in place. Three-dimensional imaging uses seismic data and the world's fastest computers to create images of deepwater basins; in other words, to "see" geologic structures (Figure 14-1) more than 5 kilometers below the surface. And recent advances in drilling techniques have significantly lowered production costs and increased oil and gas yields from known fields. Example: With much greater precision, oil drillers can now aim their bits at targets that are kilometers from an offshore platform. This technology helps recover oil and gas from a wide area using a minimum of offshore platforms. Related technology allowed companies to drill horizontal wells at record water depths. Using these techniques, companies have increased the ratio of discoveries to "dry holes" that is, exploratory wells that didn't hit oil. Dry holes can cost up to $15 million each!

We can estimate the amount of ultimately recoverable oil (i.e., the *total resource* that can be extracted) for a field, a country, or the Earth, based on our knowledge that oil is formed almost exclusively in sedimentary rocks, whose volumes can be calculated fairly accurately by using data from known fields.[5] Proven reserves are usually less than ultimately recoverable reserves by at least a factor of 2.[6] For the past 25 years, with only two exceptions, estimates of ultimately recoverable oil have ranged from around 1800 billion barrels to around 2400 billion barrels.

---

[5]Nebel, B.J., & R.T. Wright. 1998. *Environmental Science* (Upper Saddle River, NJ: Prentice Hall).
[6]Ibid., American Petroleum Institute (http://www.api.org/).

Question 4: A letter to the editor in the Washington Post[7] asserted, "[A]t the prevailing rate of consumption, world oil supplies will last for several centuries." What does "several" mean here? Is it clear or ambiguous? What assumption(s) can you identify in the quote?

To analyze the author's claim, let's first examine the issue of oil reserves. Refer to Table 14-1. The R/P ratio gives the number of years of oil left at present rates of production.

Question 5: List the world's greatest oil reserves in order by country.

Question 6: What was the world R/P ratio at the end of 1999?

Question 7: Which of these countries had the lowest R/P ratio?

Thus at present rates of consumption we can determine how long the proven reserves (1033.8 billion barrels) will last by simply dividing the total oil by the annual production (approximately 27 billion barrels per year).[8] Note that when we do this the resultant number is years, since "barrels" cancel out.

---

[7]Hoffman, M. Who wants to live like the Japanese. [Letter to the Editor.] *Washington Post,* June, 9, 1993.
[8]U.S. Energy Information Administration (http://www.eia.doe.gov/index.html).

This number represents the total number of years from 1999 that oil could be produced at current rates before it is used up, assuming constant present production. In reality, the depletion of a resource like oil does not follow such a simple pattern. Rather, oil production will gradually decline over a long period of time. It is useful for this analysis, however, to assume for the moment that it does follow this simple depletion pattern, that is, at the present rate of extraction and use.

Question 8: In what year then, starting from 1999, would the world run out of oil in this scenario?

Unfortunately, the number you just calculated is irrelevant because the rate of consumption is increasing (Table 14-2).

TABLE 14-2 ■ World Oil Demand[9]

Year	Consumption (million barrels per day)
1986	61.759
1987	62.999
1988	64.819
1989	65.917
1990	65.985
1991	66.577
1992	66.742
1993	67.043
1994	68.313
1995	70.193
1996	71.722
1997	73.700
1998	73.63
1999	74.73 (est.)
2000	75.86 (est.)

Question 9: What was the percentage increase from 1986 to 2000?

[9]U.S. Energy Information Administration. International oil data for crude oil production, available at http://www.eia.doe.gov.

Question 10:   What was the average annual growth rate in oil consumption for the 14-year period from 1986 to 2000? (The most accurate way to calculate the average rate of increase is to use the formula $k = (1/t)\ln(N/N_0)$; see *Using Math in Environmental Issues,* pages 15–16).

It is important to note how a tiny rate of increase can lead to such a big change in oil consumption over a long period of time. The concept of doubling time, $T = 70/R$, introduced earlier, can be applied here to illustrate growth of consumption.

Question 11:   What is the doubling time for the increase you just calculated?

Question 12:   How many millions of barrels of oil a day will the world be consuming when it is double the 2000 level?

Question 13:   To approximate the total amount of oil that was consumed over a given period, given the beginning and ending rates, first convert to barrels per year, then add the rates together, divide by 2, then multiply by the number of years. How much oil will have been consumed during that period you determined for Question 11?

Question 14: How does this compare to total oil reserves at the end of 1999 (from Table 14-1)? Discuss.

According to physicist Albert Bartlett,[10] "When consumption is rising exponentially, a doubling of the remaining resource results in only a small increase in the life expectancy of the resource."

Question 15: Author and environmental provocateur Garrett Hardin,[11] in addressing the issue of energy supply, said, "Whenever it is thought to be impossible to limit the growth of either population or desire, ***it is impossible to solve a shortage by increasing the supply***" (emphasis his). Explain why you agree or disagree with this statement. Can we ever accommodate the world's increasing demand for oil? (If you have some background in economics, cite reasons why an economist might disagree with Hardin's quotation).

CRITICAL THINKING QUESTIONS

Question 16: The mean estimate by the U.S. Geological Survey of recoverable oil in the Arctic National Wildlife Refuge (ANWR) is put at about 10.3 billion barrels without regard to price. Based on the mid-2001 U.S. annual consumption rate of about 20 million barrels of oil per day, how many months would these new fields last?

---

[10]Bartlett, A.A. 1978. Forgotten fundamentals of the energy crisis. *American Journal of Physics*, 46(9): 876–888.

[11]Hardin, G. November/December 1996. Letter to Editor. *Worldwatch Magazine, 9:*5.

Question 17:   Did we phrase Question 16 fairly and accurately? Is there another way of looking at how long oil from ANWR would supply the United States? Does it change your opinion about drilling in the ANWR?

ANWR's Reserves

The American Petroleum Institute,[12] among others, would like to open the Arctic National Wildlife Refuge (ANWR) on the North Slope to oil exploration. The U.S. Geological Survey (USGS) estimates 5.7 to 16 billion barrels ultimately might be produced. Environmental scientists are concerned about (1) the impact of oil exploration on the tundra environment, and (2) the impact of oil exploration and production on caribou and other migrating animals.

Here is a quote from a recent report produced by the California nonprofit Center for Energy Efficiency and Renewable Technologies (CEERT) on this subject: "Drilling the Arctic National Wildlife Refuge (ANWR) would be as shortsighted as burning the Mona Lisa to ward off a chill. The ANWR is an untamed wilderness and a priceless natural treasure. 95% of Alaska's North Slope has already been tapped. We shouldn't drill for oil in the ANWR—we should drill for oil under Detroit by raising CAFÉ standards. " The organization, in its report "The Hidden Costs of Oil"[13] further stated that

> . . . by increasing fuel efficiency by just 6% per year, new CAFÈ standards could easily reach 45 miles per gallon for cars in the next decade, saving more than $3.2 billion. This represents more oil than we import from Saudi Arabia, Kuwait, Qatar, Bahrain, the United Arab Emirates, national offshore oil production, and estimated production from the Arctic combined. CAFÉ standards of 45 miles per gallon for cars and 34 mpg for SUVs and light trucks are easily achievable. Ford's recent announcement that they will move quickly to increase fuel efficiency by 25% in their line of SUVs is ample evidence that the technology is available. GM, within days of Ford's announcement, launched competitive plans to increase fuel efficiency in their trucks and buses.

Question 18:   Issue 9 introduced the concept of CAFÉ standards for vehicles. Do you agree with the CEERT report's assertions? Why or why not? Cite evidence for your position.

---

[12]American Petroleum Institute (http://www.api.org/).
[13]Crude Reckoning: The Impact of petroleum on California's Health and Environment. CEERT, 2000. www.cleanpower.org.

ENVIRONMENTAL IMPACTS OF OIL REFINING

Critics in the U.S. Senate during late 2000 focused on the "lack of a U.S. energy policy," by which most meant a drop in domestic oil production and a dearth of refining capacity.

Oil refineries are one of the top sources of air pollution in the United States. They are one of the single largest stationary sources of volatile organic compounds, the primary component of urban smog. They are the fourth largest industrial source of toxic emissions and the single largest source of benzene emissions, which are carcinogenic.

Here are illustrations of adverse environmental impacts of oil refineries[14]:

■ A "Bay Area refinery, TOSCO, was charged with 16 willful violations after a 1999 explosion that killed four workers, the most violations ever alleged against a single California employer. This followed six other major TOSCO accidents."

■ "California ranked third in the country for unreported fugitive emissions. In 1997, the last year for which data were available in 2000, oil refineries released over 58 million pounds of assorted toxic air pollutants, continuing to be some of the worst offenders in the state."

■ "230 pounds per day, or almost 43 tons of MTBE (methyl tert-butyl ether) per year, discharge from oil refineries into San Francisco Bay (see Figure 14-2), and almost 600 pounds per day, or 110 tons, are discharged from refineries into Santa Monica Bay."

Question 19: In view of these environmental impacts, would it make more "environmental sense" to develop additional petroleum supplies by increasing energy efficiency rather than by increasing supply which must be refined? Explain your answer and list any assumption(s) you used to answer this question.

Question 20: China became a net oil importer in 1995, and the Chinese demand for petroleum is expected to increase significantly in the decades ahead, especially if the Chinese projection of a domestic market of 300 million cars is realized (see Issue 10). What are the implications for the world oil price if Chinese demand dramatically increases?

---

[14]Ibid.

FOR FURTHER STUDY

- Do you think most Americans are aware or believe that our oil reserves are becoming depleted at a rate that will exhaust them in the twenty-first century? Conduct a survey that includes this question along with other relevant questions about energy policy, mass transit, alternative energy sources, and so forth. Analyze your answers. To what do you attribute your findings?

- Ecological economists contend that when hidden costs are factored into the cost of energy, the actual price increases substantially. Determine what these hidden costs might be by consulting an excellent article by Hubbard,[15] and also a paper by Roodman.[16] Also, read the entire CEERT report.[17] Then list the sources of these hidden costs, and discuss whether consumers should pay them, and, if so, what effect that might have on energy use.

- Go to the following website http://www.worldwatch.org/alerts/000928.html and summarize the Worldwatch author's position on oil prices and consumption. Is it reasonable in your view? Justify your answer with evidence.

- Research the feasibility of utilizing shale oil. State whether you believe it is a viable replacement for oil in conventional rock reservoirs and explain your reasoning.

- Assess the impact of low prices on gasoline demand in the U.S. by preparing a report using the following data.

The Cost of Regular-Grade Gasoline at the Pump From 1940–95

Year	Price per U.S. Gallon ($)*	Year	Price per U.S. Gallon ($)*
1940	1.60	1970	1.25
1945	1.30	1975	1.45
1950	1.35	1980	2.15
1955	1.50	1985	1.30
1960	1.40	1990	1.35
1965	1.40	1995	1.20

*Price given in 1993 U.S. dollars, rounded to the nearest 5 cents.
(SOURCE: U.S. Bureau of Labor Statistics, available from http://stats.bls.gov)

Premium Gasoline Prices, Including Taxes, by Country (1994 Data)

Country	Price per U.S. Gallon ($)*	Country	Price per U.S. Gallon ($)*
U.S.	1.24	India	2.28
Australia	1.74	Japan	4.14
Brazil	1.88	U.K.	2.86
France	3.31	Venezuela	0.18
Germany	3.34		

*Price given in 1994 U.S. dollars.
(SOURCE: U.S. Bureau of the Census. 1996. Statistical Abstract of the U.S.: 1996, 116th ed., Washington, D.C.)

Breakdown of the cost of a gallon of gasoline:

---

[15]Hubbard, H.M. April, 1991. The real cost of energy. *Scientific American, 264*:36–42.
[16]Roodman, D.M. 1996. Paying the piper: Subsidies, politics, and the Environment. Worldwatch Institute paper 133. Worldwatch Institute, Washington, DC.
[17]http://www.cleanpower.org/petro/index.html.

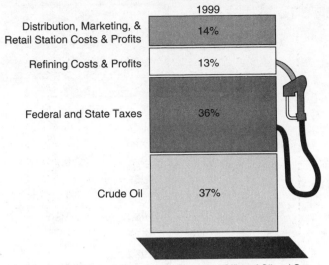

Source: Energy Information Administration Office of Oil and Gas

■ What is the typical price of gasoline today? Use the data below and plot the yearly and inflation-adjusted prices for leaded (1949–1975) and unleaded regular (1976–1999) gasoline on the axes provided.

Year	Nominal Cost (dollars)	Adjusted Cost (1996 dollars)
1949	0.27	1.55
1950	0.27	1.54
1951	0.27	1.45
1952	0.27	1.44
1953	0.29	1.49
1954	0.29	1.49
1955	0.29	1.47
1956	0.30	1.46
1957	0.31	1.47
1958	0.30	1.41
1959	0.31	1.39
1960	0.31	1.40
1961	0.31	1.37
1962	0.31	1.35
1963	0.30	1.32
1964	0.30	1.30
1965	0.31	1.31
1966	0.32	1.31
1967	0.33	1.32
1968	0.34	1.28
1969	0.35	1.26
1970	0.36	1.23
1971	0.36	1.19
1972	0.36	1.14

Year	Nominal Cost (dollars)	Adjusted Cost (1996 dollars)
1973	0.39	1.16
1974	0.53	1.45
1975	0.57	1.42
1976	0.61	1.45
1977	0.66	1.46
1978	0.67	1.39
1979	0.90	1.73
1980	1.25	2.18
1981	1.38	2.21
1982	1.30	1.96
1983	1.24	1.80
1984	1.21	1.70
1985	1.20	1.63
1986	0.93	1.23
1987	0.95	1.22
1988	0.95	1.18
1989	1.02	1.23
1990	1.16	1.35
1991	1.14	1.27
1992	1.13	1.23
1993	1.11	1.18
1994	1.11	1.16
1995	1.15	1.17
1996	1.23	1.23
1997	1.23	1.21
1998	1.06	1.03
1999	1.17	1.11

(Source: http://www.eia.doe.gov/pub/
energy.overview/aer1999/txt/aer0522.txt)

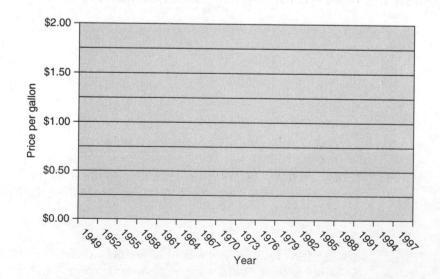

BRINGING THE WORLD TO THE U.S. STANDARD OF LIVING

KEY QUESTIONS

- What would be the environmental impact of bringing world oil consumption to U.S. levels?
- What is the relationship between U.S. oil consumption and our standard of living?

BACKGROUND AND ANALYSIS

Development agencies such as the World Bank have a stated goal of enabling "developing" nations to reach a standard of living comparable to western countries such as the United States. In this discussion, we will investigate what that might involve and what kind of impact it would have on the world's oil supply.

For purposes of comparison, a country's standard of living may be estimated by dividing the total annual value (in dollars) of goods and services (the gross domestic product, or GDP) by a country's population to derive the average per capita GDP. For example, to do this calculation for the United States, take the 1999 GDP ($8,875.8 billion)[1] and divide it by the population (approximately 276 million).[2]

Question 1: What was the U.S. per capita GDP for 1999?

Global product, the combined value of goods and services of all the world's nations, is harder to measure, but according to the World Resources Institute[3] it was approximately $26,544 billion in 1992.

---

[1]Bureau of Economic Analysis (http://www.bea.doc.gov/briefrm/tables/ebr1.htm).
[2]U.S. Census Bureau (http://www.census.gov/population/estimates/nation/popclockest.txt).
[3]World Resources Institute. 1997. *World Resources 1996–97: A Guide to the Global Environment* (New York: Oxford University Press).

Question 2:   Given a global population of approximately 5.2 billion, calculate the gross global per capita product (GGP) for 1992.

Question 3:   By what factor (2x, 4x, etc.) would the GGP have to be increased to equal the U.S. GDP in 1999?

The U.S. economy is heavily dependent on oil (see Issue 14). Oil provides roughly half of the total U.S. energy supply and virtually all of its transportation fuel. According to the Energy Information Administration (EIA), the transportation sector consumes about 28% of all domestic energy.[4] In 2000, global oil production was nearly 80 million barrels per day, and the United States alone consumed nearly 20 million barrels per day.

Question 4:   What percentage of the world oil production for 2000 was consumed by the United States?

To move the goods and services that we Americans envision when we think of a high standard of living requires oil and lots of it. To produce these goods also requires a great deal of oil. Oil, for example, is essential to American agriculture. According to Pimentel[5] and others, "the intensive farming technologies of developed countries use massive amounts of fossil energy for fertilizers, pesticides, irrigation and for machines as a substitute for human labor." And most of this fossil energy comes from oil. So even if we cannot directly correlate oil consumption with standard of living, our way of life, which we are exporting to the rest of the world, is critically dependent on oil.

However, standard of living is much more than some threshold level of median or average income, or the level of material consumption (the nature of the "living standard"

---

[4]U.S. Energy Information Administration (http://www.eia.doe.gov/oiaf/aeo/pdf/0383(2001).pdf).
[5]Pimentel, D., X. Huang, A. Cordova, & M. Pimentel. 1997. Impact of Population Growth on Food Supplies and Environment. *Population and Environment, 19:* 11.

will be considered in later questions). Most thoughtful observers prefer to use *quality of life* as a more reasonable measure of living standard instead. Measuring quality of life consists largely of evaluating intangibles such as clean drinking water and unpolluted air; access to quiet and solitude, especially in densely populated cities; alternatives to automobiles and clogged roads in making transportation decisions; access to quality medical care at reasonable prices; a job that pays a "living wage"; a safe, low-crime environment; and a relatively secure old age. Nevertheless, we can get some appreciation of what would be required to allow the world's population to enjoy U.S. lifestyles by analyzing petroleum consumption. Therefore, we will assume that a reasonable measure of our standard of living, aspired to by much of the developing world, is our per capita oil consumption.

Question 5: Do you think this assumption is reasonable? Defend your conclusion with evidence.

Question 6: Let's analyze the consequences of bringing the world's per capita oil consumption up to U.S. levels. Using the data on U.S. oil consumption given above and a 2000 population of 281.4 million,[6] calculate the U.S. per capita oil consumption in 2000 (i.e., barrels per person per year). How many gallons is that?

Question 7: To calculate world oil consumption at U.S. levels, simply multiply the U.S. per capita consumption figure by the mid-2000 world population (6.1 billion).

---

[6]U.S. Bureau of the Census (http://www.census.gov/main/www/cen2000.html).

Question 8:   World total petroleum production in 2000 was approximately 29 billion barrels. For the world to consume oil at the rate of the United States, world production would have to increase by what factor (2x, 3x, etc.) over 2000 levels?

Question 9:   Next, let's assume that the U.S. consumption were to drop by a drastic and revolutionary 50%, to roughly that of the European Union (Western Europe). In this new scenario, world oil production would have to increase by what factor (50%, 2x, 3x, etc.) to bring the world to the per capita consumption levels of Western Europe?

Question 10:   Given a 2000 world population growth rate of 1.26% per year,[7] how long would it take the world's population to double? By what year would this doubling occur? (See *Using Math in Environmental Issues,* pages 15–16 for a review of doubling time.)

---

[7]U.S. Census Bureau (http://www.census.gov/ipc/www/worldpop.html).

Question 11:   Multiply this new, doubled world population by the per capita consumption figure from Question 6. How much oil must be produced annually by the year derived in Question 10?

Question 12:   Known petroleum reserves are slightly over 1000 billion barrels. Ultimate reserves may exceed 2000 billion barrels (see Issue 14). Compare the annual oil demand of a doubled world population consuming at U.S. levels (the answer from Question 11) to the 2000 proven oil reserves of 1000 billion barrels and to probable ultimate reserves of 2000 billion barrels.

Question 13:   Evaluate the impact on air pollution if that were to occur (see Issue 17).

FOR FURTHER STUDY

■ Petroleum is transported internationally by only two ways, tanker (ship) or pipeline. A substantial increase in world oil consumption will involve major increases in tanker traffic, larger tankers, newer or larger pipelines, or a combination of the above. Research the issue, then answer the following questions.
1. Assuming that the increased oil transportation would be by tanker and that tanker size did not increase (i.e., tanker capacity remains the same), by what factor would tanker traffic have to increase?
2. Most of the world's oil available for export comes from enormous oil fields in the Middle East, specifically from Saudi Arabia, Kuwait, Iraq, Iran, and the

United Arab Emirates. Evaluate the geopolitical impact of an increase in oil tanker shipments from this region in the light of the past 50 years of conflict in the region.

■ Tankers leaving the Persian Gulf, where the extra oil production is likely to occur, must navigate the northwest Indian Ocean. This area is subject to tropical storms, averaging more than five per year. It is also the site of numerous spectacular coral reef communities. Already the mouth of the Persian Gulf is the site of the world's highest density of ocean shipping, averaging more than 400 million tonnes of oil per year by 1996.[8] It has also been the site of at least four major tanker accidents.

1. Evaluate the environmental consequences of a large increase in tanker traffic.
2. Each year "normal" leaks and spills from tankers and terminals dump 250,000 barrels of oil into the Persian Gulf. By what factor would this increase, based on the increased rate of tanker shipments that you previously calculated, assuming that all the increase in oil production would be shipped by tanker from the Persian Gulf?

■ Another way to assess standard of living is by housing size. Many Americans are now building large houses, often with floor area of over 371.2 m^2 (4000 ft^2), although the average size of new homes is around 2200 ft^2.[9] Determine what proportion of houses are in the "larger" categories. How does this compare with the 1980 census? To other countries? What would be the environmental impact if everyone lived in these "McMansions" or "starter castles" (as they have been called)?

■ Identify as many high "quality of life" intangibles (such as clean air) as you can and explain your choices.

[8]Allen, J.L. 1997. *Student Atlas of Environmental Issues* (Guilford, CN: Dushkin/McGraw-Hill).
[9]U.S. Bureau of the Census (http://www.census.gov/prod/99pubs/99statab/sec25.pdf).

RENEWABLE ENERGY: IS THE ANSWER BLOWING IN THE WIND?

KEY QUESTIONS

- What are the major sustainable sources of energy?
- What is their actual and potential role in U.S. and global energy production?

Michal Moore, a member of the California Energy Commission, authored the following report:[1]

> California opened its huge $23 billion electricity market to competition two years ago. Though the jury is still out on when consumers will save money, one thing has become remarkably clear, . . . the vast majority of residential customers that have switched power suppliers have . . . purchased cleaner electricity generated by a variety of renewable energy resources.
>
> Over 90 percent of the 200,000 consumers that have switched power suppliers have voted for the environment with their green electricity purchases. In fact, demand for green power is surpassing expectations. Applications for rebates offered by the California Energy Commission to lower the cost of green power for smaller consumers have outstripped projections of available funds.
>
> There are a number of other surprises that emerge after two years of customer choice:
>
> - Over 30 churches have switched their own facilities to green power and are encouraging their parishioners to switch to green power under a program entitled Episcopal Power & Light.
> - Approximately 30 local governments have also switched to green power. The city of Santa Monica is currently the world's largest all-renewable city, but Oakland is considering a purchase that would put it in the global lead.
> - Over two dozen companies—including Kinko's, Patagonia, Fetzer Winery, Birkenstock, MCI/WorldCom, Toyota Motor Sales, Time Warner, Real Goods, Lucky Brand Dungarees, and Kaiser Permanente—have also switched to a variety of green power.
>
> According to the California Public Utilities Commission, trends in state consumer purchases of green power continue to go up. The green power market grew by 15 percent each quarter of 2000—with a 25 percent jump in the last quarter.
>
> California's landmark restructuring law, Assembly Bill 1890, was passed in 1996. Since restructuring began, private sector power plant developers are now moving forward to build 53 new renewable projects in California. These renewable projects will generate over 500 MW of new renewable capacity.

---

[1]http://www.cleanpower.org/news/moore.htm. Michal Moore was a prime architect of the state's current renewable energy programs and policies.

There has never been a renewable energy program as successful as the auction held by the California Energy Commission that led to the building of all these new green power projects. These new state-of-the-art renewable energy projects, developed with just $162 million in public funds, helped create new power plant projects that represent over $700 million in total capital investment. The cost of providing this new clean power was significantly below predictions, and Californians will benefit from a diversity of clean power projects, ranging from wind to geothermal to biomass power plants, strategically located throughout the state.

These new renewable energy projects also deliver important air benefits. This new clean power capacity offsets 286,113 tons of carbon dioxide that would be emitted into the atmosphere every year (and contribute to the global climate change threat) if the same electricity were to be provided by the generic mix of power sources consumers buy if they don't switch to a clean power provider. In addition, these new power plants avoid the annual release of 1,300 tons of sulfur dioxide (acid rain) and 1,634 tons of nitrogen oxides (urban smog).

Renewable energy technologies at present include those that employ wind (Figure 16-1), solar, geothermal, biomass, and hydroelectric dams to produce power. Dams are not a truly renewable system, since siltation behind the dams renders them inoperative over a period of time ranging from several decades to one or more centuries. Moreover, hydroelectric dams usually have negative effects on rivers, since they prevent the migration of many fish species.

These renewable energy sources have the following in common:

■ They cause no direct air or water pollution since no combustion is involved.
■ They require no shipment of fuels, so environmental disasters such as the 1989 Exxon Valdez oil tanker spill in Alaska's Prince William Sound will not occur.
■ They have, however, some significant limitations:
  1. Using present technologies, all must produce electricity instead of a more "portable" fuel such as petroleum.
  2. Solar power depends on sunlight, so, lacking a viable means to store the electricity produced during the day, there must be a supplemental energy source at night.
  3. A minimum wind speed is required for wind power (depending on technology), so large-scale wind power generation is limited to sites where threshold persistent winds occur. Moreover, the cost of wind power is strongly affected by two factors, average wind speed and interest rates. Since the energy that the wind contains is a function of the cube of its speed, small differences in average

FIGURE 16-1 Wind farm in Minnesota. (©Layne Kennedy/CORBIS)

winds from site to site mean large differences in electricity production and, therefore, in cost. Even though there is no fuel cost for a wind plant, such plants are expensive to build; thus, most of the cost is in the capital required for equipment manufacturing and plant construction. This in turn means that wind's economics are highly sensitive to the interest rate charged on that capital, assuming investors borrow the money to build the plant.

4. Storage technologies for electricity that would allow these renewable energy sources to be used extensively in the transportation sector have not yet been perfected.

Despite these limitations, energy production from renewable sources has increased substantially in recent years. For example, in 1999, for the first time, the world installed more new wind generating capacity than nuclear capacity. And many scientists believe hydrogen-based systems may join the list in the twenty-first century.

ENERGY CONSUMPTION

Next, we will analyze trends in energy consumption. Energy consumption is measured in British thermal units (BTU) and quad BTU. (One BTU is approximately equal to the amount of heat necessary to raise the temperature of 454 g (1 lb) of water by 1°F, and 1 quad BTU equals 10^{15} BTU.) U.S. energy consumption increased from 66.4 to 80.2 quad BTU from 1970–88, and reached 94.2 quad BTU in 1997.[2]

Question 1: What was the percentage increase in energy consumption over the period 1970–1988? 1988–1997?

Question 2: What was the average yearly increase in energy consumption over the period 1970-97? (Use the formula $k = (1/t)\ln(N/N_0)$; see *Using Math in Environmental Issues*, pages 15–16)

[2]U.S. Bureau of the Census. 1999. Statistical Abstract of the U.S. (http://www.census.gov/prod/99pubs/99statab/sec19.pdf). U.S. Energy Information Administration. 1996. Annual Energy Outlook. Prepared by the Office of Integrated Analysis and Forecasting. U.S. Department of Energy. Washington, DC.

Question 3:   What is the doubling time of energy consumption, based on your answer to Question 2? (Use the formula T = 70/R).

Question 4:   Based on the 1970–1997 rate of increase, estimate U.S. energy use in 2010, 2020, and 2030. (Use the compound interest formula: future value = present value x $(e)^{kt}$.)

Question 5:   Independent electric power generation in the United States is growing rapidly. These firms generate electricity and sell it to utilities or industry. The data for electric power generated from solar and wind technologies from 1990 to 1997 are given in Table 16-1. What was the percentage increase in electric power generated from solar and wind systems from 1990-95?

TABLE 16-1 ■ Electric Power Generated From Solar and Wind Systems (1990-97)[3]

Year	Quad BTU
1990	0.91
1991	0.95
1992	0.98
1993	1.00
1994	1.04
1995	1.15
1996	1.23
1997	1.23

Question 6:   What was the annual rate of increase in electric power generated from solar and wind systems from 1990 to 1997?

---

[3]U.S. Bureau of the Census. Statistical Abstract of the U.S.: 1999 (http://www.census.gov/prod/99pubs/99statab/sec19.pdf).

Question 7: What is the doubling time of electricity production from solar and wind systems, based on the data for 1990 to 1997?

Question 8: Assume this rate of increase were to hold to the year 2020. What would be the amount of electricity (in quad BTU) produced from these sustainable sources in 2020? (Use future value = present value x $(e)^{kt}$.)

Question 9: How does the production figure you got in your answer to Question 8 compare with the value for 2020 you calculated in Question 4?

Question 10: To evaluate whether these sustainable sources could meet the U.S. electricity needs, you will have to consider, among other things, the limitations of solar and wind power. What are the limitations of solar and wind power? Keep in mind that we need electricity all the time (i.e., 24 hours a day) and there are few ways to store electricity once it has been produced. (In this context, it would be helpful to determine how electricity demand varies over the day. When is the daily maximum demand? minimum demand? Consult your local utility's public relations office for details.)

Question 11: U.S. energy consumption is commonly divided into sectors: industry, commercial, residential, and transportation. See Table 16-2 for the breakdown of U.S. energy consumption for 1997. Do you think renewable energy resources could replace all use of fossil fuels? Explain your reasons.

TABLE 16-2 ■ Breakdown of U.S. Energy Consumption for 1997[4]

	From All Sources (quad BTU) (%)	From Renewable Energy Sources* (quad BTU)
Total consumption	96.1 (100%)	2.8
Industry and miscellaneous	35.3 (37%)	2.2
Commercial and residential	34.7 (36%)	0.5
Transportation	26.1 (27%)	0.0

*Renewable energy sources include hydroelectric, wood and wood waste, municipal solid waste, other biomass, wind, photovoltaic, and solar sources.

AN EXAMPLE OF RENEWABLE ENERGY: WIND-GENERATED ELECTRICITY IN DENMARK

Denmark is a world leader in the production, domestic use, and exportation of wind turbines. The statistics on the growth of electricity production from wind turbines in Denmark are given in Table 16-3.

TABLE 16-3 ■ Growth of Electricity Production from Wind Turbines in Denmark[5]

Year	Electricity Consumption in Denmark (gigawatt-hours)*	Share of Wind Energy in Total Consumption (%)	Employment in Wind Power Industry
1984	23,732	0.1	
1985	25,256	0.2	
1986	26,586	0.5	
1987	27,559	0.6	
1988	27,946	1.0	
1989	28,235	1.5	
1990	28,551	2.1	
1991	29,594	2.5	
1992	30,085	3.0	3100
1993	30,625	3.4	4100
1994	31,237	3.6	6000
1995	31,491	3.7	8500
1996	32,392	3.8	N/A
1997	32,356	6.0	N/A
1998	N/A	N/A	12,000

*Gigawatt-hours = units of 1 million kWh.

---

[4]U.S. Energy Information Administration. Annual Energy Outlook (http://www.eia.doe.gov/oiaf/aeo/pdf/appa.pdf).
[5]The Danish Wind Turbine Manufacturers Association (http://www.windpower.dk/publ/stat9804.pdf).

Question 12: Calculate the average growth rate of electricity consumption in Denmark from 1985 to 1997. Now plot the growth of electricity consumption on the axes below. Discuss your findings and comment on any trends in consumption you observe.

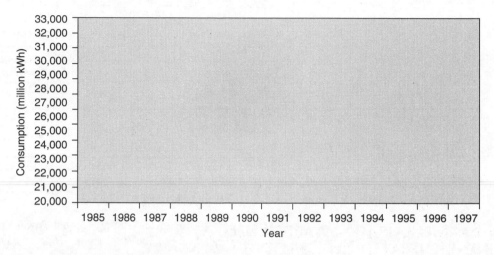

Question 13: Plot the rate of increase in electricity generated from wind on the axes below. Wind is estimated to provide 13% of Danish electricity consumption in 2000.

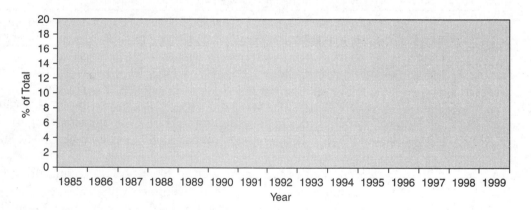

Reducing energy consumption from fossil-fueled sources is often claimed to be undesirable because it may threaten the jobs that the fossil-fuel industry supports.

Question 14: Evaluate the growth of jobs in the Danish wind turbine industry (admittedly there is little information to go on—this is, after all, a new industry). Is there cause for optimism that renewable energy sources can provide a viable employment base? Why or why not?

Table 16-4 ■ States with the Greatest Wind Potential.[6]

The Top Twenty States for wind energy potential, as measured by annual energy potential in the billions of kWhs, factoring in environmental and land use exclusions for wind class of 3 and higher.					
1	North Dakota	1,210	11	Colorado	481
2	Texas	1,190	12	New Mexico	435
3	Kansas	1,070	13	Idaho	73
4	South Dakota	1,030	14	Michigan	65
5	Montana	1,020	15	New York	62
6	Nebraska	868	16	Illinois	61
7	Wyoming	747	17	California	59
8	Oklahoma	725	18	Wisconsin	58
9	Minnesota	657	19	Maine	56
10	Iowa	551	20	Missouri	52

WIND ENERGY DEVELOPMENT IN THE UNITED STATES

Oct. 11, 2000—Cinergy Corp. subsidiary Cinergy Global Power Ltd. said yesterday that it and Sea West Wind Power Inc. had put the 28-turbine, 16.8 megawatt Foote Creek IV Wind Power, Wyoming wind farm into commercial operation. The project is located on a rim of the Medicine Bow Range of the Rocky Mountains, northwest of Laramie. It is among the windiest spots in the nation. (See Table 16-4).

"One of (our) key strategies is to invest in renewable energy projects worldwide," said John Bryant, president of Cinergy Global Resources. Sea West has developed three large wind projects on the Medicine Bow Range and, with completion of Foote Creek IV, the rim will have a total installed capacity of 84.75 megawatts.[7]

FOCUS: WIND ENERGY IN HAWAII

Hawaii, alone among the 50 states, cannot call upon its neighbors to provide it with electricity. As a result, the state of Hawaii is actively supporting the development of a mix of renewable energy resources including solar power, biomass, hydropower, wind power, geothermal energy, and ocean thermal energy conversion. Imported oil is used to supply about 90% of Hawaii's energy needs. No place else in the United States is so critically dependent on imported oil.[8]

Recent studies have attempted to identify the potential for wind energy throughout Hawaii. Here is a conclusion of a government-sponsored report on the feasibility of wind and solar energy: "Wind and solar technologies have more recently entered the commercial market and, although the power output from these sources tends to be intermittent, their energy potential remains largely untapped." As of 1999, the state obtained less than 2 MW of electricity from wind, but an additional 40 MW of capacity was being constructed.

Here is an assessment of the advantages of wind.[9]

1. Employment benefits would occur because construction and operation of renewable energy projects generate more jobs than does a comparably sized fossil fuel

---

[6]http://www.energyonline.com/Restructuring/news_reports/news/1011ren.html.
[7]www.awea.org, website of the American Wind Energy Association.
[8]http://www.hawaii.gov/dbedt/ert/wind_hi.html#anchor1613678.
[9]http://www.hawaii.gov/dbedt/ert/hes3/index.html.

plant; to the extent that the investment in these plants came from out of state, the economy would show an improvement.

2. A greater use of Hawaii's abundant indigenous renewable energy resources would help to insulate the state from fossil fuel price escalation and supply disruptions.
3. Money spent on indigenous energy would largely remain in the state.

There are also substantial environmental advantages to using renewable energy. Replacing growth in energy use with renewable as opposed to fossil fuels could reduce oil spill risks from increased imports of crude oil into the state.

But there are also limitations, mainly centered on the high cost of land on many of the islands, which could mitigate against use of the land for wind systems, as well as the nature of the state's electricity distribution system. Here is the state's assessment of the overall potential.[10]

> Renewable energy projects can theoretically provide all the new generation required to satisfy projected electricity demand increases in the state through 2014. However, a number of factors influence the ability of these resources to meet these new energy demands. In Maui County, supplying future increases in electrical energy demand could be accomplished with renewable projects that are cost effective even under the most conservative assumptions. In Hawaii and Kauai Counties, this could be accomplished with projects that are cost competitive under nominal scenarios. If conservative assumptions were used, Hawaii and Kauai could still obtain 50% and 25%, respectively, of their projected energy demand growth from renewable energy projects. On Oahu, under nominal assumptions, renewable energy projects could meet over 30% of projected electricity demand increases, while under optimistic conditions, they could provide all of the electricity required to meet projected demand.

The phrase *nominal assumptions* usually refers to a reasonable set of assumptions of interest cost of loans, land lease costs, and so on. Although the state assumes that wind projects could displace agriculture and thus might not be viable, many wind sites are operated on sites compatible with agricultural activity. However, in Hawaii, the two main crops, sugar and pineapples, require harvesting techniques that are not as compatible with wind energy as other crops. Although there is no technical incompatibility with these crops, landowners and plantation managers have indicated that it may make their traditional harvesting techniques more difficult. With pineapple cultivation, for example, the location of wind turbine towers could greatly interfere with the equipment used to harvest pineapples. The burning of cane fields may also affect wind turbines.[11]

A further complication to developing wind-energy sites could be local or state zoning regulations, but in Hawaii, this would not be a significant drawback.

One factor that may ultimately decide the issue could be the future price of oil, which is the source of 90% of Hawaii's energy, as you saw above.

FOR FURTHER STUDY

- Go to http://www.hawaii.gov/dbedt/ert/hes3/plan.html#section01 and assess the potential for wind development on the "Big Island" of Hawaii. How do the trade winds and the positions of the volcanoes on the island affect the wind potential?
- Although most petroleum is used to fuel motor vehicles, some would argue that oil is too valuable to burn in vehicles or in power plants. Research the importance of oil to the petrochemicals industry and make a list of products produced from petro-

---

[10]http://www.hawaii.gov/dbedt/ert/hes3/index.html.
[11]http://www.hawaii.gov/dbedt/ert/hes3/plan.html#section01.

chemicals. Consider specifically how oil is used in plastics manufacture, in medicine, and in agriculture. Do you feel petroleum is more valuable in petrochemical applications than as a fuel for transportation? Explain your reasoning. Is this a good reason to encourage development of renewable energy sources for the transportation sector? Why or why not?

■ Solar photovoltaic technology employs another renewable energy source. Research the growth of the photovoltaic industry by searching the World Wide Web, using the terms "solar" and "photovoltaics." And access the website of British Petroleum, which has one of the world's biggest investments in photovoltaics (www.bp.com). Discuss its potential over the next several decades.

■ Call your electricity supplier and find out what proportion of their energy is produced from sustainable sources. Ask what plans they have to increase this amount.

■ Electric cars at present are fueled by batteries, and the current generation is not very efficient. Research recent developments in battery technology. Are there new solutions to this efficiency problem? Identify problems with using batteries in the transportation sector.

■ The deregulation of the electric power industry in California has begun after a fashion (see the beginning of this Issue). Consumers can choose their electricity provider much as they can now choose their long-distance telephone company. Research the environmental implications of deregulation by contacting the California Energy Commission's website at http://www.energy.ca.gov.

■ For details on the economics and technologies needed to develop wind energy, go to www.awea.org. Go to http://www.awea.org/projects/index.html for information on U.S. wind development state by state. Choose your own state and evaluate the extent to which its wind energy is being used for electricity. If your state does not have any wind developed, find out why. What sources produce your state's electricity? Coal? Nuclear? Gas?

■ How "predictable" is the future oil price? Go to www.eia.gov and determine how the government projects the world price of oil. Some have argued that oil producing countries could sabotage the development of renewable energy by dropping the price low enough periodically to render renewables "uncompetitive" with oil. Is this a reasonable possibility? Why or why not?

Issue 17

ENERGY WASTE AND AVOIDABLE AIR POLLUTION: GETTING THE PICTURE ABOUT VCRS, TVS, AND ICED DRINKS

KEY QUESTIONS

- How much pollution results from burning fossil fuels?
- What is the cumulative impact of some seemingly insignificant sources of energy use?
- Can we cut pollution by becoming more energy-efficient?

Energy availability and prices took center stage in the United States at the beginning of 2001. In California, the state was struggling through electricity shortages, caused in part by unprecedented increases in demand (demand *longages*). Throughout the country, consumers were faced with the highest natural gas prices in history, and in the New England states, residents who heated their homes with oil faced another winter of high fuel oil prices. During the past summer, U.S. drivers complained of high gas and diesel prices.

What's going on? What factors are behind these new energy "crises"? Was it excessively strict environmental regulations, as some claimed? Were restrictions on oil and gas drilling at fault, as many Western politicians asserted? Or could energy waste and poor education about energy fundamentals share the blame? In this issue we will assess areas in the U.S. economy where energy waste can be measured and evaluated.

AIR POLLUTION FROM FOSSIL-FUELED ELECTRIC POWER SOURCES

It is often asserted that the United States, and indeed the entire world, must and will increase energy use indefinitely, regardless of the logical and physical impossibility of such activity. Yet the sheer waste of energy in modern societies goes largely unappreciated. Many studies have shown that aggressive, cost-effective energy conservation strategies could render additional power plant construction unnecessary if there were also a simultaneous curtailment of population growth.

As of 2001, the transportation sector ran almost exclusively on gasoline and diesel fuel, and over 50% of America's electricity was generated by coal.[1] Moreover, regardless of its source, energy that is wasted still generates avoidable impacts, affecting other segments of society (such as the healthcare system, forests, and agriculture; see *For Further Study* below).

---

[1]U.S. Energy Information Administration. 2001. Annual Energy Outlook. Prepared by the Office of Integrated Analysis and Forecasting. U.S. Department of Energy, Washington, DC. (http://www.eia.doe.gov/oiaf/aeo/index.html).

The U.S. Environmental Protection Agency's Green Lights Program[2] estimates that for each 454 g (1 lb) of coal burned to produce electricity (without emission controls), the following are released[3]:

- 1.1 kg (2.4 lbs) of CO_2
- 4 g of NOx
- 10 g of SOx

Burning fuel oil to produce electricity emits:

- 900 g of CO_2 per kilowatt-hour (kWh) of electricity produced
- 1.9 g of NOx per kWh
- 5.4 g of SOx per kWh

Natural gas is viewed as a "clean" alternative fuel to produce electricity, but it emits:

- 590 g of CO_2 per kWh of electricity produced
- 2.1 g of NOx per kWh
- 0 g of SOx per kWh

Electricity used unnecessarily is also responsible for the release of toxic heavy metals into the atmosphere and hydrosphere as a by-product of fossil-fuel combustion. Among the most toxic of these are mercury, nickel, chromium, and cadmium. The average emission factors for these metals are as follows:

- 0.04 mg of mercury (Hg) per kWh of electricity produced
- 1.18 mg of nickel (Ni)
- 1.3 mg of chromium (Cr)
- 0.05 mg of cadmium (Cd)

An average 500 megawatt, coal-fired power plant (Figure 17-1) emits about 100 kg of arsenic per year, 2 kg of cadmium, 50 kg of lead, 450 metric tonnes of particulates (see below), 4500 tonnes of sulfur oxides (SOx), and 9000 tonnes of nitrogen oxides (NOx).[4] Sulfur and nitrogen oxides in the atmosphere are primary sources of acid precipitation. Finally, the plant produces nearly 1 million tons of carbon, mainly as carbon dioxide, a key greenhouse gas.[5]

So, to avoid having to deal with unnecessary pollution, it is clearly in our long-term interest to avoid having to produce electricity if it will only be wasted or used inefficiently. Moreover, since we now import well over half of our oil, it is important to avoid unnecessary oil use as well. This is far from the case, however.

In the following sections we will evaluate and try to quantify two small examples of wasted electricity and in the *For Further Study* discussion we will consider energy waste and ways to increase efficiency.

---

[2]U.S. Environmental Protection Agency. Green Lights Program (http://www.epa.gov/greenlights.html/).
[3]Although the numbers in the coal example add up to more than 454 g (1 lb), this is not a violation of the law of conservation of matter, since oxygen in the atmosphere combines with carbon, sulfur, and nitrogen during combustion to produce these compounds.
[4]American Lung Association (http://www.lungusa.org).
[5]Mathews, J. False negative on the environment. *Washington Post*, November 23, 1996.

FIGURE 17-1 Aerial view of the Navajo power plant, a coal-fired utility. Pollution from this power plant often contributes to haze over the Grand Canyon. (©Dewitt Jones/ CORBIS)

Energy Waste

Example 1: The Video Cassette Recorder

According to the U.S. Bureau of the Census.[6] 88% of U.S. households own at least one video cassette recorder (VCR), but 20% cannot program the clock. In other words, in 20% of these households the symbol "12:00" is constantly flashing on and off, or the clock tells incorrect time. Let us assume that 100 million VCRs are owned,[7] 20 million by people who can't set the clock (that's 20% of 100 million). Let's further assume that a reasonable 2 watts per hour of electricity are required to run the clock, 24 hours a day, 365 days a year. Recall that VCRs have no "off" switch: that is, it is impossible to turn the clock off without unplugging the device, even if the VCR is off.

Question 1: What is the amount of electricity required each year to run these 20 million flashing clocks? Express your answer in kWh per year.

---

[6]U.S. Bureau of the Census. 2000. Statistical Abstract of the U.S.: 1999, Washington, D.C. (http://www.census.gov/prod/www/statistical-abstract-us.html).
[7]There are also many VCRs in organizations such as government offices, schools, colleges, and universities that are not considered in our calculations.

Question 2:   Let's calculate the capacity (or size) of a coal-fired power plant that would be needed to supply the wasted energy. First, divide the total kWh used by the flashing clocks by 8760, the number of hours in a year. (A typical fossil-fueled power plant produces electricity at a rate ranging from 300,000 to 1,000,000 kilowatts per hour. The plant is able to operate around three-fourths of the time, due to periodic shutdowns for routine maintenance.) Next, multiply the kilowatts/hour by 4/3 to get the capacity of the plant. Then add 10% to cover the amount of electricity needed to remove SOx.

Such a plant, if fueled by coal, would cost around $300 million to build (not including interest). The electricity from coal could be produced at around 4.4 cents per kWh, assuming some environmental costs were included (other than those associated with CO_2 emissions). The electricity would be sold to residential users at around 8 cents per kWh.

Question 3:   What is the annual cost to run the clock on each VCR?

Question 4:   What is the annual cost to run all the blinking VCR clocks in the United States?

Approximately 1 kg of SOx, which contributes to acid precipitation, is produced for each 100 kWh of electricity, if the power plant does not use SOx scrubbers, and as of early 2001 many did not. Thus, to the dollar cost from Question 4 must be added the cost of the secondary effects and pollution (air pollution, injuries and lung disease occurring in miners, coal dust pollution along rail lines, etc.) from a coal-fired power plant operated unnecessarily.

Example 2: Iced Drinks

A drink purchased at a convenience store or at a restaurant, especially a "fast food" restaurant, automatically comes with anywhere from 100 to 200 g of ice, and the practice is almost unconscious. Iced water and other iced drinks literally permeate our society all year round, even in the dead of winter. They could be called more American than apple pie. Yet the environmental consequences of the American penchant for iced drinks is rarely evaluated. In this issue we will consider some of the implications of our preference for iced drinks year round.

To convert 1 kg of liquid water at 0°C to 1 kg of ice at 0°C requires 80 kilocalories (Cal). More energy is needed to further reduce the temperature of the ice, which is usually kept near −20°C. Therefore, we estimate the total energy required to produce ice at 150 Cal per kg.

Question 5: Using a conversion of 1 Cal = 0.00116 kWh of electricity, how much electricity is needed to turn 1 kg of water at 0°C to ice at 0°C?

Question 6: Assume an airliner with 200 passengers uses around 250 g of ice per person during a flight from Washington, DC to San Francisco. How much electricity is consumed solely to make the ice?

Question 7: There were 8.2 million scheduled aircraft departures carrying 598.9 million passengers in the United States in 1997.[8] Assume that every passenger receives iced drinks with 250 g total ice (enough for two drinks). Assess the environmental benefit if they all gave up their ice. Would this make a dent in the U.S. consumption of electricity of over 3100 billion kWh?[9]

Example 3: Instant-on TVs

Televisions (and indeed many electronic home appliances such as computers and printers) have an "instant-on" feature: that is, part of the television's components continuously draw electricity, whether the TV is on or off. Typical TVs draw 40 watts of electricity when turned off.[10] Let's see if that can help explain California's energy crisis. The population of California is approximately 34 million as of 2001.[11] In the United States, there

[8]U.S. Bureau of the Census. 2000. Statistical Abstract of the U.S.: 1999, Washington, DC. (http://www.census.gov/prod/www/statistical-abstract-us.html).

[9]U.S. Bureau of the Census. 2000. Statistical Abstract of the U.S.: 1999, Washington, DC. (http://www.census.gov/prod/99pubs/99statab/sec19.pdf).

[10]http://www.lbl.gov/Science-Articles/Archive/leaking-watts.html.

[11]U.S. Bureau of the Census. 2000. Statictical Abstract of the U.S.: 1999, Washington, DC. (http://www.census.gov/prod/www/statistical-abstract-us.html).

were an average of 2.4 televisions per home as of 1997,[12] so we will take that as a reasonable value for California as well. There were approximately 2.6 persons per household.[13]

Question 8: Use the above information to estimate the number of households in California as of 1998 to 3 significant digits.

Question 9: How many personal TVs were owned by residents?

Question 10: At 40 watts per hour to power the TVs when turned off, what is the electricity use?

At the end of 2000, officials and utility experts pointed to a "shortage" of about 1100 MW.

Question 11: Could California solve its short-term electricity "crisis" simply by unplugging its TVs when not is use?

Recall that this analysis does not take into consideration the hundreds of thousands of TVs in hotel rooms, airports, and other public places in the state.

[12]U.S. Bureau of the Census. 2000. Statistical Abstract of the U.S.: 1999, Washington, DC. (http://www.census.gov/prod/99pubs/99statab/sec18.pdf).

[13]U.S. Bureau of the Census. 2000. Statistical Abstract of the U.S.: 1999. Washington, DC. (http://www.census.gov/prod/99pubs/99statab/sec01.pdf).

SUMMARY QUESTIONS

Question 12: Do we really need all the ice we use? Discuss whether it is worth the air pollution generated by the fossil fuel burned, recalling that over 50% of the electricity used in the United States is produced by burning coal.

Question 13: You may have concluded that ice production is an infinitesimal source of U.S. energy consumption. Assume 1000 industries were responsible for polluting a river. Could they each argue logically that since each one was responsible for such a small part of the problem, that no one should be held accountable? Justify your answer.

Question 14: Evaluate the significance of these exercises and activities, given the magnitude of U.S. energy use. Are they relevant to the discussion of energy waste and avoidable pollution? Why or why not?

Question 15: As you have seen, VCRs, TVs, and many other home electronic devices continue to draw power even after the appliance is turned off. The reason manufacturers give is that consumers demand that the devices "warm up" quickly. Do you think that a real "on/off switch" to save electricity is a good idea, based on your answers to Questions 1 to 4 above? Would you be willing to wait the additional 15-30 seconds needed to warm up the device?

FOR FURTHER STUDY

■ Your Family's VCRs: How many VCRs does your family own? TVs? Calculate your family's monthly energy bill from the VCR clocks and TVs, assuming 2 watts for the VCR clocks and 40 watts for the TVs when they are "off." How many digital clocks are there in your house?

- Your Family's Energy Use: Go to http://www3.crpud.net/enersrv/usage.asp and assess your personal or family's energy use.
- Your Electricity Supplier: Obtain a copy of the latest annual report of your local electricity supplier. They usually have public relations offices with toll-free numbers that you can call to get a free copy. Look up the proportion of electricity produced by your supplier that comes from coal. Find the SOx emissions per kWh or per 100 kWh of electricity produced. If it isn't published in the annual report, call the company's Public Relations office and find out how you can obtain the information. Most utilities (i.e., electricity suppliers) have a Clean Air Compliance Division (or similarly named office) as well. Find out how the company encourages energy efficiency, if at all. Many companies have rebate programs, low-interest loans, and so on available for their customers. Has your family requested an energy audit in your home?
- Group/Individual Research Projects: **Energy use at your institution.** Most people as individuals are somewhat aware of their personal energy use and make a least some effort to avoid unnecessary consumption. But when we put on our "organization" hats and go to work or class, we usually undergo a radical change and don't care or become uncomfortable about accepting responsibility. In addition, in most organizations no one is "in charge" of energy conservation. Consider the following examples from a typical institution; you could without doubt identify additional examples from your own surroundings.

  1. **Water coolers.** As a part of a plan to rebuild the District of Columbia's troubled school system, the D.C. School Board plans to renovate several schools. The renovation plan calls for, among other things, the installation of eight refrigerated water coolers at an installation cost of $1000 each for each school. Most public and commercial buildings in this country have several of these. They can't be shut off, and many can't even be unplugged! Therefore, they operate throughout the year. In winter, the water is often too cold to drink. Much of the water we drink from the coolers is simply reflex action: The person drinking is simply taking a few sips of water out of habit, and most of the water trickles into the drain. Thus, much of the energy to refrigerate the water is simply wasted. Finally, the quality of the water may be suspect.

     Activity: Count the number of refrigerated water coolers in an academic or administrative building on your campus. Somewhere on the back of the cooler is printed the watts per hour the cooler uses. The coolers are usually affixed to the wall, making it impossible to read this information. Obtain this information from your institution's physical plant office. Or sleuth around the machine (be careful if there are exposed wires and water pipes!) and find out the manufacturer's name and perhaps the model number. If you can, call the manufacturer and, if you can determine the cooler's model number, find out the energy consumption of the cooler.

     Armed with your consumption figures, monitor the cooler for about 15 to 30 minutes and record how much of that time the cooler is on. They generally have thermostats that shut them off when the water reaches a sufficiently low temperature. Then figure out the maximum energy use of the cooler per month. Determine how many coolers there are in your building or in all of the buildings on campus. Calculate the approximate energy use and cost per month. Do we need all this chilled water? Explain your reasoning.

  2. **Lighting.** Since no one is "responsible," lights in college and university classrooms are often left on all day and sometimes through the night. Most classrooms use 34- or 40-watt fluorescent bulbs (or you can ask your physical plant manager for the wattage used at your institution). There are often several bulbs

per fixture. Count the number of light bulbs in a classroom and calculate the daily energy use. Assuming a cost of 8 cents per kWh, calculate the daily cost. Find out from your administration whether the lights are turned out at night. Are there signs in the rooms asking users to turn off lights when not in use?

3. **Energy Audits.** Ask your institution's administration how much it spends annually on electricity and find out what energy conservation or efficiency programs it operates. Divide the annual electricity bill by the number of students and figure out the per capita cost, since you pay at least part of the bill. At an average electricity cost of 8 cents per kWh, how much does it cost to run the lights for 1 hour in one of your classrooms? What is the cost to leave the lights on in one room, often unused, from 5 PM until 7 AM the next day? How many similar rooms are there in your institution?

Year	Ft^2(million)	Total Energy Cost (million $)	Total Kbtu (X 10^6)	kBTU/ft^2
1996	25.2	31.6	2,948.5	116.95
1997	26.0	33.0	3,493.0	137.67
1998	27.2	33.2	3,326.4	140.06
1999	27.6	33.9	3,791.7	138.88

Here are statistics for South Carolina colleges with housing for 1996–1999[14] Plot square feet, total energy cost, and total energy used in kBTU on the same graph. Are the slopes of the lines the same? Which is growing faster, if any?

Total electricity costs for South Carolina colleges were $33.936 million in 1999. The state paid an average of 4.7 cents/kwh that year. For natural gas, the state paid an average of 36.5 cents per therm, which is half the price for natural gas at the end of 2000. What do you forecast will happen to the State's natural gas bill for 2001 compared to 1999 and its electric bill between 2000 and 2010?

4. **Lighting and Safety.** Do you believe that buildings are safer if all of the lights are on? Is there a potential security cost in energy conservation? If so, do you believe that money spent on crime prevention (e.g., to provide people with good educations and the opportunity for decent jobs) could pay an energy dividend? To what extent should the money we spend on lighting be counted as part of the hidden "cost" of crime?

■ Visit the EPA website and get information on the Green Lights Program,[15] which seeks to improve energy efficiency through lighting efficiencies in buildings. EPA reports that up to 40% of electricity used in lighting buildings can be saved by installing new lighting systems that may pay for themselves in less than 6 months! Find out what types of high-efficiency lights are available.

■ Visit the website of the Sustainable Campus Policy Bank[16] (SEAC) and get information on campus audit programs. Find out if your campus has had an energy audit done, and if not, urge them to do so.

■ Visit the website of the Sierra Club[17] and the Sacramento (California) Municipal Utility District (SMUD)[18] and get information on energy waste and energy efficiency. They usually make their press releases available on-line, as well as sum-

---

[14]Energy Use in SC's Public Facilities, 1999. SC Energy Office.
[15]U.S. Environmental Protection Agency. Green Lights Program (http://www.epa.gov/greenlights.html/).
[16]http://iisd1.iisd.ca/educate/policybank.asp.
[17]http://www.sierraclub.org/.
[18]http://www.smud.org/.

maries of their research reports. How much energy could the United States save if we applied all feasible cost-effective efficiency measures?

■ What impact does low price for energy have on energy efficiency? Research this subject and write a report focusing on how much gasoline could be saved if higher prices were charged, for example.

■ Pollution generated by burning fossil fuels has an adverse impact upon other segments of society: Buildings are damaged by "acid precipitation," agricultural production is decreased by exposing plants to air pollution, and chronic respiratory problems such as asthma are worsened by air pollution. Research this issue by consulting the websites of the American Lung Association[19] Sierra Club, Worldwatch Institute,[20] and so on.

■ Can you identify other acts of nearly unconscious consumption or waste that seem to serve little useful purpose? Select one and try to quantify its impact.

■ The next time you travel by air, ask airline personnel how much ice they take on board for each leg of the flight. Compare it to the numbers we have presented here.

■ Go to: www3.crpud.net/enersrv/usage.asp and examine the Home Appliance Energy Use Guide. Assess your personal energy use based on this list.

---

[19]American Lung Association (http://www.lungusa.org).
[20]http://www.worldwatch.org/.

GREENHOUSE GASES, GLOBAL CO_2 EMISSIONS, AND GLOBAL WARMING[1]

KEY QUESTIONS

- What is the composition of the earth's atmosphere?
- What processes have influenced the atmosphere's composition?
- How does the earth's atmosphere interact with the ocean?
- What are greenhouse gases?
- What impacts of global climate change could be associated with greenhouse gas increases?

Earth's atmosphere is a relatively thin shell. About 95% of it is contained within 14 km (8.6 mi) of the earth's surface. All life on earth depends on the atmosphere, but despite this knowledge, humans are altering the atmosphere's composition. Scientific evidence has confirmed that emissions from the burning of fossil fuels (see Figure 17-1), from industrial sources such as cement manufacture, and from deforestation have changed and continue to change the makeup of our atmosphere. In addition, trace gases such as methane and chlorofluorocarbons (CFCs and HCFCs) are having an impact on the atmosphere wholly out of proportion to their concentration.

THE EARTH'S EARLY ATMOSPHERES

Our current atmosphere is thought to be the earth's third over its cosmic history. The first was comprised of light gases, primarily hydrogen (H_2) and helium (He_2). Slightly heavier inert gases such as argon (Ar_2), krypton (Kr_2), and neon (Ne_2) were also likely present, along with dust. This first atmosphere was blown away as the sun reached critical mass and began its internal fusion reactions, and by heat generated by the early molten earth between 4 and 5 billion years ago.

The second atmosphere was formed between 3.8 and 4 billion years ago when gases escaped from the earth as its crust solidified. In addition, the hypothesis that bombardment by comets contributed gases to the earth's second atmosphere is gaining acceptance by scientists. This second atmosphere contained a variety of gases but was devoid of oxygen, a gas necessary to support most of the earth's present life forms.

---

[1]Dr. Eric Koepfler of Coastal Carolina University contributed to this Issue.

THE DEVELOPMENT OF THE CURRENT ATMOSPHERE

The earth's third and current atmosphere was produced partly by the metabolism of living organisms. Photosynthetic organisms, the cyanobacteria, appeared by 3.8 billion years ago and began to produce oxygen as a by-product. Over the next 2 billion years, atmospheric O$_2$ concentrations rose and CO$_2$ concentrations fell as a direct result of photosynthesis, as well as the sequestration (storage, or "fixing") of carbonate rock (limestone and dolostone: CaCO$_3$ and CaMg(CO$_3$)$_2$). This increase in atmospheric oxygen then set the stage for two major evolutionary events on the planet: the evolution of aerobic (oxygen-using) life forms and the formation of the *ozone layer*.

Increases in levels of oxygen in the hydrosphere and atmosphere poisoned sensitive microorganisms and led to the evolution of new microorganisms capable of more efficient respiration: using oxygen to liberate energy from organic compounds. Today, natural anaerobic microbial communities are restricted in their habitats to Mid-Ocean Ridge hydrothermal systems, deep marginal or isolated seas, organic-rich sediments, hot springs, and regions beneath the earth's surface. The development of aerobic metabolism also permitted the evolution of multicellular organisms, which required more energy to support their increased biomass. All existing multicellular life forms employ aerobic metabolism.

Increases in atmospheric oxygen concentration eventually triggered the formation of the layer of ozone (O$_3$) that now exists in the upper atmosphere. Although considered a pollutant at ground level, the ozone layer serves as an important filter of harmful high-energy ultraviolet radiation, which can cause skin cancer in humans.

Prior to the existence of the ozone layer, the earth's terrestrial and ocean surface were exposed to extremely high intensities of ultraviolet radiation. The surface ocean waters filtered out some of this radiation and thus provided some protection to organisms, but it is likely that ocean primary production (that is, production of high-energy compounds from photosynthesis) was still limited by the high ultraviolet light intensities. The terrestrial surface fared worse and was possibly sterilized by this radiation: The first widely accepted evidence for land plants, for example, appears in rocks that are only about 400 million years old.

THE ATMOSPHERE'S CURRENT COMPOSITION

The present-day atmosphere is composed primarily of N$_2$ gas (78.08% by volume) and oxygen (20.94% by volume). Also present, in quantities less than 1% by volume, are, in order: Ar, H$_2$O, CO$_2$, Ne, He, CH$_4$, SOx, NO$_2$, CO, NH$_3$, and O$_3$. The major controls upon the composition of the atmosphere and the cycling of these compounds are interactions with the Earth's biosphere (living matter) and lithosphere (rock and geological processes such as volcanism). Presently, O$_2$ levels seem to be stable, but CO$_2$ levels are not and display seasonal and longer-term trends. Seasonal changes in CO$_2$ concentration are related to primary production changes due to changing light durations. Longer-term (decade, century) increases in CO$_2$ are due to a variety of anthropogenic (human-caused) inputs as well as changes in land use that reduce the ability of terrestrial biota to absorb CO$_2$.

Because the amount of CO$_2$ in the atmosphere is very small, the concentration is easily changed by the addition of CO$_2$ from various sources.

FUNCTIONS OF THE ATMOSPHERE

In addition to providing the oxygen needed by most of earth's life forms, the atmosphere provides a significant thermal insulation, preventing extreme changes in temperature over the daily light-dark cycle. Unequal heating of the earth's atmosphere and terrestrial surface create long-term climate and short-term weather patterns. The winds that result from these heating differences and resultant pressure differences also drive ocean currents. The atmosphere also transfers vast quantities of heat from equatorial to polar latitudes.

CHANGES CAUSED BY HUMANS

After decades of research, scientists have finally concluded that humans have changed the composition of the earth's atmosphere. Before the industrial revolution (ca. 1750), clear-cutting of forests in Europe, China, and the Middle East, and later in North America, set the stage for modifying the atmosphere's composition. Cutting and burning forests liberates CO_2 in two ways: Carbon from the vegetation is converted to CO_2, and soils, devoid of their tree cover, emit CO_2 at much greater rates than before.

Since the beginning of the industrial revolution, atmospheric concentrations of the following greenhouse gases have increased: carbon dioxide by 30%; methane concentrations have more than doubled; and nitrous oxide concentrations have risen by about 15% (Table 18-1). Greenhouse gases allow short wavelength radiation from the sun to pass through the atmosphere, but they absorb the longer wavelength radiation (i.e., infrared) that is emitted by the Earth. In the past 2 ½ centuries, a global human population that has grown tenfold; the need for energy to support industrial development, to heat homes, cook food, watch television, and surf the Internet; as well as the increased use of automobiles, have resulted in the burning of great stores of fossil fuel.

Fossil fuels, including coal, oil, and natural gas, were formed by the preservation and slow anaerobic decomposition of ancient plant and phytoplankton deposits. These deposits took tens of millions of years to form but we are now burning them at a vastly more rapid rate. A key by-product of fossil fuel consumption is CO_2. Since the industrial revolution, we have added CO_2 to the atmosphere more rapidly than it can be absorbed by its variety of sinks. This has led to a slow but steady increase in CO_2 concentration that will result in at least a doubling of pre-1860 atmospheric CO_2 content by the year 2150, if present trends continue.

Methane is another greenhouse gas whose concentration has become elevated because of human activities. It is emitted by cows, and flooded farmlands (i.e., rice paddies), as

Table 18-1 ■ Changes in the Global Concentration of Greenhouse Gases Since the Preindustrial Period[2]

	CO_2	CH_4	N_2O
Preindustrial concentration	280 ppmv	700 ppbv	275 ppbv
Concentration in 1994	358 ppmv	1720 ppbv	312 ppbv[2]
Rate of concentration change[1]	1.5 ppmv/yr	10 ppbv/yr	0.8 ppbv/yr
Atmospheric lifetime (years)	50–200[a]	12[b]	120

ppmv = part per million by volume; ppbv = part per billion by volume
[1]Concentration increases in CO_2, CH_4, and N_2O are averaged over the decade beginning in 1984.
[2]Estimated from 1992–1993 data.
[a]No single lifetime for CO_2 can be defined because of the different rates of uptake by different processes.
[b]Defined as an adjustment time that takes into account the indirect effects of methane on its own lifetime.

[2]Intergovernmental Panel on Climate Change (IPCC), 1995. Climate Change 1995: The science of climate change: Contribution of Working Group I to the Second Assessment Report of the Intergovernmental Panel on Climate Change. J.T. Houghton, L.G. Meira Filho, B.A. Callander, N. Harris, A. Kattenberg & K. Maskell, eds. (Cambridge University Press: New York).

well as municipal landfills. All of these have increased dramatically within the last two centuries.

FOCUS: ENVIRONMENTAL IMPACTS OF CLIMATE CHANGE

At its meeting in Shanghai in January 2001, the Intergovernmental Panel on Climate Change (IPCC) issued its most comprehensive report to date on the environmental implications of climate change. Over 150 delegates from nearly 100 governments met from January 17–20 to consider the Working Group I contribution to the Third Assessment Report of the IPCC "Climate Change 2001: The Scientific Basis."[3] The full report, which runs to over 1000 pages, is the work of 123 lead authors, who in turn drew on more than 516 contributing authors. The report went through extensive review by experts and governments. In line with IPCC Principles and Procedures, after line-by-line consideration, the governments unanimously approved the Summary for Policymakers of the report and accepted the full report.

Here are the major conclusions of this latest report:

■ Since the IPCC's 1995 Report, confidence in the ability of models to project future climate has increased. Reconstructions of climate data for the past 1000 years, as well as model estimates of natural climate variations, suggest that, while uncertainties remain, there is an anthropogenic "signal" in the climate record of the last 35-50 years; that is, this change has certainly been affected by human activities.

■ Continuing analyses of information from tree rings, corals, ice cores and historical records for the Northern Hemisphere indicate that the increase in temperature in the 20th century is likely to have been the largest of any century during the past 1000 years. It is likely that the 1990s were the warmest decade and 1998 was the warmest year.

■ In the mid- and high-latitudes of the Northern Hemisphere, it is very likely that snow cover has decreased by about 10% since the late 1960s, and the annual duration of lake- and river-ice cover has shortened by about two weeks over the 20th century. It is likely that there has been about a 40% decline in Arctic sea-ice thickness during late summer to early autumn in recent decades.

■ Since 1750, the atmospheric concentration of carbon dioxide has increased by 31% from 280 parts per million (ppm) to about 367 ppm today. *The present CO₂ concentration has not been exceeded during the past 420,000 years and likely not during the past 20 million years* (our emphasis).

■ The globally averaged surface temperature is projected to increase by 1.4–5.8°C from 1990 to 2100. This is higher than the 1995 Second Assessment Report's projection of 1–3.5°C, largely because future sulfur dioxide emissions (which help to cool the Earth) are now expected to be lower. This future warming is on top of a 0.6°C increase since 1861.

■ Global average water vapor concentration and precipitation are projected to increase. More intense precipitation events are likely over many Northern Hemisphere mid- to high-latitude land areas. The observed intensities and frequencies of tropical and extra-tropical cyclones and severe local storms, however, currently show no clear long-term trends, although data are often sparse and inadequate.

■ Sea-levels are projected to rise by 0.09 to 0.88 meters from 1990 to 2100. Despite higher temperature projections these sea level projections are slightly lower than

---

[3]http://www.ipcc.ch/.

the range projected earlier (0.13 to 0.94 meters), primarily due to the use of improved models, which give a smaller contribution from glaciers and ice sheets.

What will the impacts be of global climate change associated with greenhouse gas increases? Examine Figure 18-1 for an overview.

Human health will be directly affected by increases in the *heat index,* which can impact healthy people, but especially affects the elderly and those with heart and respiratory illnesses. Higher temperatures will also increase ozone pollution in the lower atmosphere, which is a threat to people with respiratory illnesses. Global warming may also increase the incidence of some infectious diseases, particularly those that usually appear only in warm areas. Diseases that are spread by mosquitoes and other insects, including malaria, dengue fever, yellow fever, and encephalitis, could become more prevalent if warmer temperatures and wetter climates enable those insects to become established farther north. Already, municipalities are planning increased spraying of insecticides to combat tropical vector-borne (mainly by mosquitoes) diseases.

Another important impact of global climate change will be alterations in precipitation patterns. Predicted changes in climate are expected to enhance both evaporation and precipitation in most areas of the United States. The net balance of these processes influences the availability and quality of water resources. In areas expected to become more arid, like California, lower river flows and lower lake levels could impair navigation, reduce hydroelectric power generation, decrease water quality, and reduce the supplies of water available for agricultural, residential, and industrial uses. In other areas, increased precipitation, which is expected to be more concentrated in large storms as temperatures rise, could increase the incidence of flooding.

Shifts poleward in climatic regimes would also affect the types of crops that can be grown. Studies conducted in the 1980s generally concluded that climate change would have severe impacts on agriculture. More recent assessments have suggested that these might be partially offset, at least in the United States by longer growing seasons and enhanced production from higher CO₂.

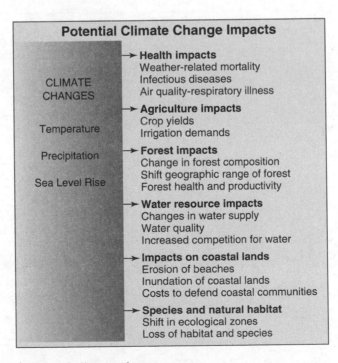

FIGURE 18-1 Impacts of climate change.

Climatic change will cause a shift in natural community composition as plant and animal species migrate to maintain their preferred habitats. For example, the projected average 2°C (3.6°F) warming that will occur this next century could shift the ideal range for many North American forest species by about 300 km (200 mi) to the north. Many plant species lack seed dispersal mechanisms that can adjust this rapidly, and thus these species may be susceptible to regional extinction. Coastal wetland plant communities may lose habitat because they may not be able to keep up with the predicted enhanced rate of sea level rise, or the migration paths of these communities may be blocked by human development.

Among wildlife, certain birds and fishes are expected to be greatly impacted by predicted climate change. A large number of duck species are dependent upon "prairie potholes" found in the Northern Great Plains. A drier climate would decrease the amount of open water ponds in this region, with a commensurate reduction in duck populations. Among fish, those that inhabit inland aquatic environments are expected to be more vulnerable than coastal or marine species. Lake-locked fishes have little recourse in seeking cooler waters. Fish that inhabit north-south rivers may be able to migrate northward to seek cooler water, but fish in east-west oriented rivers and lakes will not be able to escape warming impacts. The results of a 1995 EPA study, "Ecological Impacts From Climate Change: An Economic Analysis of Freshwater Recreational Fishing," suggest that the overall diversity of fishes in U.S. rivers and streams is likely to decline because of the loss of cold water forms.

CLIMATE CHANGE AND THE NORTH ATLANTIC CIRCULATION

In a number of scientific papers in such journals as *Nature, Science,* and *GSA Today,* Columbia University Geoscientist W.S. Broecker has warned of "an oceanic flip-flop." Global warming, it is argued, could interrupt the system called the *thermohaline circulation.* Thermohaline circulation maintains the flow of the Gulf Stream, that great current that brings vast quantities of heat in the form of warm, salty water from the tropics that warms much of Europe (Figure 18-2). The entire system seems to depend on rapid, in-

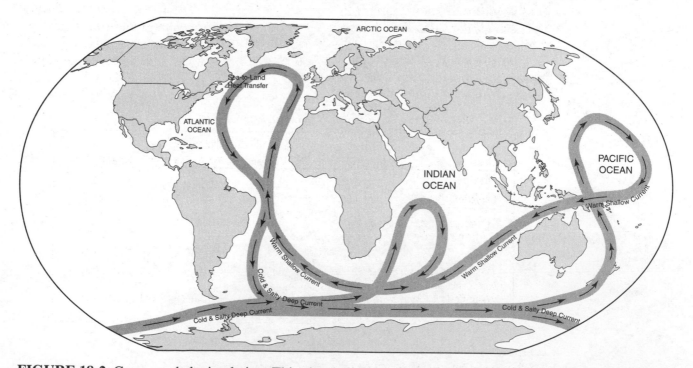

FIGURE 18-2 Conveyor-belt circulation. This circulation is initiated when cool North Atlantic water sinks. It then flows southward and flows into the Indian and Pacific Oceans.

tense cooling of hypersaline (extra-salty) water off Greenland. If global warming caused increased rainfall, reduced wind speeds over the North Atlantic, or led to the melting of freshwater glaciers in Greenland, salt concentrations in surface waters could fall, leading to less mixing of surface and deep waters. This would interrupt the flow of the Gulf Stream, bringing a cooler climate to northern Europe. More recently, Broecker has suggested that these effects could turn off the deep-ocean conveyor belt completely, possibly triggering an ice age. Evidence for such flip-flops has been found in geological records obtained from ice cores and deep-sea sediments. Of particular concern is the fact that these events have occurred over time periods as short as *four years*. Broecker refers to the oceans as the Achilles' heel of the climate system.[4]

Question 1:   In the 1999 paper in GSA TODAY,[5] Broecker says, ". . . an extreme scenario is an unlikely one, for models suggest that in order to force a conveyor shutdown, Earth would have to undergo a 4 to 5 degree C warming." Based on this statement and the 2001 IPCC report, comment on the possibility of a conveyor shutdown in the North Atlantic before 2100.

FACTS ABOUT THE ATMOSPHERE

To analyze the effects of human activity on the atmosphere there are some facts you need to know.

Half the mass of the atmosphere is within 5 km (3.1 mi) of the surface, 80% of the mass is within 15 km (9.3 mi) of the surface, and 90% is within 20 km (12.4 mi). Twenty kilometers is not very far; it's about a 10- to 11-minute drive on most interstate highways. A breakdown of the major gases that comprise the Earth's atmosphere is given in Table 18-2. The proportions of the gases (ignoring water vapor, which is variable) in the atmosphere remain virtually constant to a height of 25 km.

TABLE 18-2 ■ Major Gas Composition of the Atmosphere*

Gas	Volume (%)	Molecular Weight (approximate)	Weight (%)
Nitrogen	78.000	28	75.00
Oxygen	21.000	32	23.00
Argon	00.930	40	1.30
Carbon dioxide	00.035	44	0.05
Average atmosphere	100.000	29	100.00

*Percentages do not add up to 100 due to rounding, approximation, and ignoring water vapor and all minor and trace gases.
Weight (%) is found by multiplying the volume (%) by the molecular weight of that gas and dividing by the average molecular weight of the atmosphere.

[4]UN Environmental Programme, GEO 2000: Global Environmental Outlook (http://www.unep.org/Geo2000/).
[5]Broecker, W.S. 1999. What if the conveyor were to shut down? *GSA TODAY*, 9 (1): 2–7.

WEIGHING THE ATMOSPHERE

At sea level, the weight of a 1 cm^2 column of atmospheric air is approximately 1 kg. This value, 1 kg/cm^2, is also known as the atmospheric pressure. For this analysis, we will assume a featureless Earth without continents. This will simplify, but not significantly alter, your calculations.

Question 2: How much does the atmosphere weigh over each square meter of surface? Express this in tonnes (units of 1000 kg). Recall the conversion factors: 10^4 cm^2 = 1 m^2; 1 tonne = 1000 kg = 10^6 g.

Question 3: How much does the atmosphere weigh over each square kilometer (km^2)? Recall that 10^6 m^2 = 1 km^2.

Question 4: The area of the Earth's surface is approximately 516 $\times$ 10^6 km^2. What is the weight of the entire atmosphere in tonnes? Express your answer in scientific notation, in units of 10^{12} tonnes.

It is worth pointing out that this is by no means an incomprehensibly large number. The mass of the Earth is 6 $\times$ 10^{27} g. How does the atmosphere's mass compare with that of the Earth?

CARBON DIOXIDE IN THE ATMOSPHERE

You have just calculated the weight of the entire atmosphere. Now let's figure out how much of this weight is CO$_2$.

Question 5: CO_2 comprises 0.05% (1/20 of 1%) by weight of the atmosphere (see Table 18-1). What is the weight of CO_2 in tonnes in the atmosphere?

Question 6: World CO_2 emission from deforestation and burning of fossil fuels is roughly 7.7 billion tonnes per year.[6] What percentage of 2580 billion (2580×10^9) tonnes (the total amount of CO_2 in the atmosphere) is 7.7 billion (7.7×10^9) tonnes?

Question 7: Thus, according to these calculations, we are adding CO_2 to the atmosphere now at the rate of approximately _____% per year. At this rate, can we account for the observed increase in CO_2 in the atmosphere since the 1950s? CO_2 content in 1959 was measured at 316 ppm (parts per million), and was 368 ppm in 1999.[7] What is the percentage increase in CO_2 over this 40-year period?

Question 8: Is this percentage increase within the order of magnitude suggested by your calculations? (Order of magnitude means roughly within the same power of 10; e.g., the ratios 3.5 cm/yr and 6.5 cm/yr are within the same order of magnitude, whereas 3.5 cm/yr and 26.8 cm/yr are not.)

[6]Worldwatch Institute. 2000. *Vital Signs 2000* (New York: W.W. Norton and Co.).
[7]Ibid.

Question 9: Recall the introduction to this Issue: Do you think the CO_2 produced by human activities stays in the atmosphere? If not, where would it go? Identify a major "sink" or storehouse for this anthropogenic (human-generated) CO_2.

Question 10: Sulfur dioxide is mainly produced by burning fossil fuels coal and oil. It is a toxic chemical in high concentrations and is a major contributor to acid precipitation, which has been documented to adversely affect human health, buildings, and crop yields, among other things. Would you favor relaxing pollution reduction regulations to help reduce global warming, as some have proposed? Use evidence to support your answer.

SEA LEVEL CHANGES AND GLOBAL WARMING

It has been suggested that global warming will result in rising sea level, as polar regions warm and polar ice melts. This has been questioned, since some assert that warming the poles will result in more snow, thus more accumulation of ice, and not melting. There is, however, one impact of warming that can be easily evaluated arithmetically, and that is the impact on sea level resulting from warming the water.

The coefficient of thermal expansion of seawater is approximately 0.00019 per degree Celsius. This simply means that, given a volume of sea water, that volume will expand by this fraction as a result of heating the water, per degree. Now since ocean basins are constrained by their bottoms and "sides," the only way to go is up. Note that this also could result in extensive coastal flooding.

To calculate how much a temperature increase would increase sea level, simply multiply the average ocean depth (in cm) by the coefficient of thermal expansion (c.t.e.) by the number of degrees of temperature rise. (Note that we put a 1 before the c.t.e. value to obtain the height of sea level after the warming. If you simply multiply the average depth by the c.t.e. you will get the number of centimeters that sea level will rise.)

Question 11: With this understood, how much would each 1°C increase in sea water temperature cause sea level to rise? Express your answer in cm (30 cm = 1ft). Recall the average depth is 3,800 m (3.8 km).

FOR FURTHER STUDY

- Research the "carbon cycle" using Nebel and Wright[8] or another text. Discuss the importance of limestone and fossil fuels as carbon storehouses (sinks).
- Many countries with sizable populations living in coastal areas, as well as some small countries with their entire land area near or at sea level, are understandably acutely concerned about the impact of sea-level rise associated with human alteration of the atmosphere. The Maldives are one such country, and Bangladesh is discussed elsewhere in this book. Research the position of the government of the Maldives on sea-level rise and write a paragraph or two summarizing it. What responsibilities do countries which produce the preponderance of greenhouse gases have to mitigate the impact on countries like the Maldives? What assumptions did you use to answer this question?

---

[8]Nebel, B.J., & R.T. Wright. 1998. *Environmental Science* (Upper Saddle River, NJ: Prentice Hall).

ESTUARINE POLLUTION: CHESAPEAKE BAY

KEY QUESTIONS

- What are the sources and the impacts of pollution on Chesapeake Bay and its watershed?
- Who is responsible for managing the impact of human activity on Chesapeake Bay?
- What actions are being taken to reduce the environmental impact of population growth there?

BACKGROUND

We will consider three important aspects of Chesapeake Bay: the Bay proper, the Bay's watershed[1] (Figure 19-1), and the Bay's airshed[2] (Figure 19-2). The Chesapeake Bay, which formed 8000 years ago as rising sea level drowned the mouth of the Susquehanna River, is the nation's largest and most productive estuary.[3] Surveys have shown, however, that most residents of the Bay's watershed are unaware of that fact, and most do not even realize that they live within the watershed. To increase public awareness, in 1997 the states in the watershed put signs along major highways, indicating the watershed's boundaries. The watershed's 64,000 mi^2 area includes parts of six states and the District of Columbia and includes 1653 local governments. The watershed had a human population approaching 16 million at the end of 2000[4] and this population was growing at around 1.2% per year. The airshed measures 418,000 square miles, or roughly 6 1/2 times the size of the Bay's watershed.

Question 1: What is the doubling time (see *Using Math in Environmental Issues,* pages 15–16) of the watershed's population?

---

[1]The watershed is the area drained by streams that feed the Bay.
[2]The Bay's (or any location's) airshed is the geographic area that is the source of airborne pollutants that can affect the Bay (or that particular location).
[3]An estuary is a coastal embayment where freshwater from rivers or ground water mixes with saltwater.
[4]http://www.census.gov/.

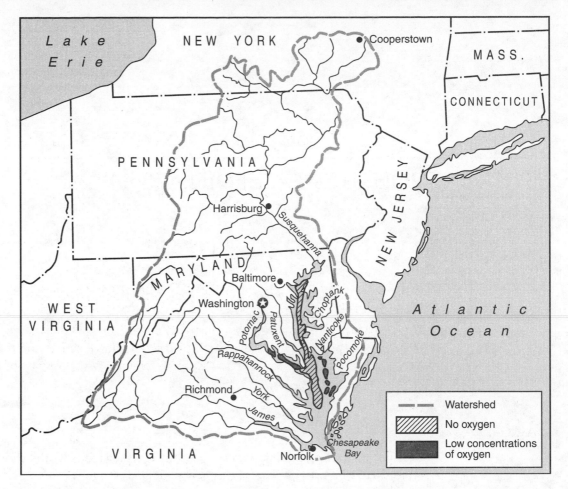

FIGURE 19-1 Chesapeake Bay and its watershed. (From Blatt, H. 1997. *Our Geologic Environment.* Upper Saddle River, NJ: Prentice Hall)

Question 2: How many states and Canadian Provinces are included in the Bay's airshed?

ENVIRONMENTAL RESEARCH ON CHESAPEAKE BAY

In 1983, the U.S. Environmental Protection Agency (EPA) published the first of what was to become a series of studies of Chesapeake Bay's environment and concluded that it was seriously threatened by *nutrient enrichment*. Nutrients are substances that are essential for plant growth, including compounds of nitrogen and phosphorus. The EPA also expressed concern regarding

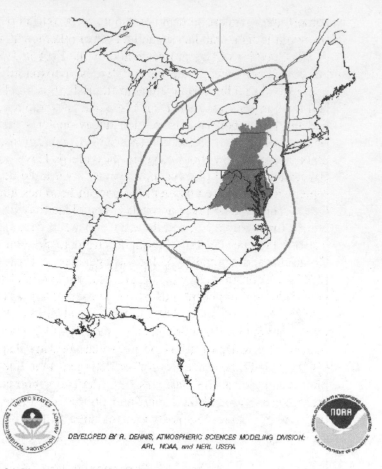

DEVELOPED BY R. DENNIS, ATMOSPHERIC SCIENCES MODELING DIVISION;
ARL, NOAA, and NERL USEPA.

FIGURE 19-2 Chesapeake Bay's airshed. (NOAA)

1. The overharvesting of oysters, crabs, and fish.
2. The 2000 dams and other obstructions to fish passage that had been constructed on the Bay's tributaries over three centuries.
3. The toxic substances entering the Bay, mainly from industrial and commercial sources.[5]

By 1991, the Bay's oyster population had nearly disappeared. On the plus side, a ban on rockfish (also known as striped bass) harvesting apparently allowed that species to recover. This recovery, however, brings its own problem: the decline of prey species. Bay anchovies are suffering from overpredation due to the increase in the numbers of striped bass, which are feeding heavily on the bay anchovy. Why? Because populations of Atlantic menhaden, the striper's main prey, have been reduced due to pollution, dams, and overfishing. As you can see, a change in one part in an ecosystem can have unexpected changes throughout.

Bay grasses had covered most of the Bay's shallow bottom (its average depth is only 7 m) to a depth of about 2 m until the 1950s, but by 1983 their area had been reduced by 90%. Bay bottom grasses (also called submerged aquatic vegetation (SAV)) may be the key to a healthy Chesapeake Bay. The grasses provide oxygen for the water, food for

---

[5]McConnell, R.L. 1995. The human population carrying capacity of the Chesapeake Bay watershed: A preliminary analysis. *Population and Environment, 16* (4): 335–351.

some Bay organisms, attachment spots for a myriad of tiny organisms, and hiding places for juvenile fish, crab larvae, and a host of other small animals.

The decline of Bay grasses noted by the EPA in 1983 was caused largely by the nutrient enrichment mentioned earlier. Excess phosphorus, and to a lesser extent nitrogen, cause floating microscopic algae to flourish, thus blocking sunlight to the shallow bottom and the bottom grasses. The surface algae grow rapidly, quickly deplete the nutrients, die, and fall to the bottom. There, their decay depletes the water of oxygen, resulting in hypoxic (low oxygen) or anoxic (no oxygen) conditions. Since most bottom-dwelling animals cannot live without oxygen, they die or leave, and the bottom, now without plant cover and depleted of oxygen, becomes a veritable desert. Thus, without Bay grasses to replenish the depleted oxygen, the bottom becomes inhospitable to most Bay animal life. Other changes take place as well: Green algae may be replaced by cyanobacteria and diatoms, for example (for details, see Cooper and Brush).[6]

In 1987, the states of Maryland, Virginia, and Pennsylvania joined with the District of Columbia and a number of federal agencies to form the Chesapeake Bay Commission (CBC) to coordinate Bay restoration and protection efforts. One of the CBC's goals was a 40% reduction in phosphorus by the year 2000, a goal that it was unable to meet, although progress had been made. The Commission also set annual numerical goals for phosphorus reduction: 8.43 million pounds per year. By 1992, phosphorus had been reduced by 16%, mainly as a result of a ban on phosphate-bearing detergents that took effect in January 1989, over the heated objection of detergent manufacturers. In 1992, the average phosphorus content in the Bay was 0.03 mg/l. For comparison, streams that drain forested areas contain around 0.02 mg/l phosphorus and those that drain pasture contain around 0.05 mg/l; a representative value for streams that drain urban areas is 0.075 mg/l and those that drain farmland contain about 0.15 mg/l phosphorus.

Question 3: Calculate the volume of Chesapeake Bay. The Bay's surface area is roughly 11,000 km² and, as mentioned earlier, the average depth is 7 m. Express your answer in cubic kilometers, cubic meters, and liters.

Question 4: How much phosphorus was dissolved in the Bay in 1992? Express your answer in kilograms.

---

[6]Cooper, S.R., & G.S. Brush. 1991. Long-term history of Chesapeake Bay anoxia. *Science, 254:* 992–996.

THE IMPACT OF SEWAGE ON CHESAPEAKE BAY

A consultant's report estimated the sewage flow from all sources in Virginia's portion of Chesapeake Bay's watershed to be 522 million gallons per day (MGD; 1 gal = 3.8 L) by the year 2000.

Question 5: Based on this rate of sewage flow, how long would it take for Chesapeake Bay to fill up with treated sewage, assuming the sewage wasn't flushed to the ocean?

Question 6: Based on a 1.2% population growth rate, and assuming sewage flow is proportional to population growth, how long would it be before 1 year's sewage discharge from Virginia equals the entire volume of Chesapeake Bay?

Question 7: As part of Chesapeake Bay protection efforts, Virginia required that all municipal and industrial sewage treatment plants, and all industries that emit phosphorus, reduce the phosphorus concentration in sewage outflow to a level of 2 mg/l by the year 2000. Based on attainment of that standard, how much phosphorus did Virginia sewage treatment plants discharge into the Bay in 2000? Express your answer in kilograms and pounds per year.

Question 8: At 2 mg/l, how does sewage runoff compare to the typical values for streams given above?

Question 9:   Do you think a 2 mg/l standard will be sufficient to clean up the Bay? State your reasons. What additional information would you like to have in order to answer this question more thoroughly?

Question 10:   A 1995 study of the Cameron Run Watershed in Fairfax County, Virginia suggests that the amount of phosphorus discharged by streams that drain urban areas has been significantly underestimated. Researchers found the phosphorus content in these streams to be 2.5 times greater than land-use models had assumed. Discuss the implications for Bay protection in the light of the rapid urbanization of the region you noted in Issues 5 and 6.

New Developments in Chesapeake Bay Research

Recent research suggests that between 20% and 35% of nitrogen in Chesapeake Bay is from air pollution. A third of this air pollution is from cars, power plants, and farm fields in the watershed itself, but as much as two-thirds is from power plant emissions in Ohio, Kentucky, Michigan, and other states as far away as Alabama[7] (see Figure 19-2) This is apparently due to three factors:

1. Prevailing winds that blow from the southwest, northwest, and west.
2. The lack of control of nitrogen pollution on the part of midwestern coal-burning utilities.
3. Smokestacks from the plants that emit the pollution high enough into the atmosphere so that it can be carried over 800 km before settling out!

Accordingly, to address water and air pollution problems in the Bay's watershed and throughout the eastern United States, the U.S. EPA has ordered 22 states in an arc from Massachusetts to Missouri to reduce pollutants, including nitrogen (in the form of oxides of nitrogen), or lose federal highway funds. The affected states had until 1999 to devise action plans and will have until 2005 to actually reduce emissions. The states most heavily affected and the percentage of nitrogen oxide reductions required are: West Virginia

---

[7]EPA orders 22 states to reduce nitrogen oxide emissions. *Washington Post*, October 11, 1997.

(44%), Ohio (43%), Missouri (43%), Indiana (42%), Kentucky (40%), Illinois (38%), Alabama (36%), Wisconsin (35%), Tennessee (35%), and Georgia (35%).

Question 11: To what extent should the people in these states be held accountable for air pollution carried beyond their boundaries? Include your reasons and explain them.

These nitrogen oxides not only contribute to the pollution of Chesapeake Bay, they also contribute to acid precipitation. For example, in the early 1990s, northern Virginia, Washington DC, and its Maryland suburbs had the most acidic rainfall of any urban area in the United States, according to the U.S. Geological Survey. And air pollution resulting from excessive nitrogen oxides has been implicated in respiratory problems, especially in children and elderly people.

Many allergists and environmental scientists say our healthcare system is unfairly burdened with extra costs to treat persons affected by air pollution from coal-burning power plants (and motor vehicles; you will also find a similar observation dealing with turfgrass in Issue 12). Some believe these costs should be paid in the form of higher electric rates by those who choose coal-fired power plants and who decline to control emissions from these plants.

Question 12: Do you agree that the extra costs imposed by using coal to generate electricity should be borne by the consumers of the power? Explain your reasons.

PFIESTERIA AND MANURE

An outbreak of a mysterious alga of the genus *Pfiesteria* (Figure 19-3) killed tens of thousands of fish in tributaries of Chesapeake Bay in 1997. The outbreaks seemed to be limited to streams that drained areas of Maryland with huge chicken farms. These farms disposed of up to 800,000 tons of nutrient-enriched chicken manure each year by spreading it on fields, some immediately adjacent to streams. The federal government and Maryland officials proposed to stop the leaching of nutrients from the manure into the streams by paying the farmers to leave unplowed buffer strips consisting of grass or (better yet) trees along streams that drained through their fields. The grass and trees would absorb nutrients in the runoff, preventing it from contaminating the Bay. The initial cost was reported to be around $250 million.

FIGURE 19-3 Sores on menhaden associated with an outbreak of *Pfiesteria*. The incidence of harmful algal blooms in coastal and estuarine environments has been increasing worldwide recently. (J. Burkholder)

Question 13: To what extent does this $250 million represent a subsidy for chicken production and consumption? Explain your reasoning.

Question 14: To what extent, or under what circumstances, do government officials have the right to tell farmers and growers what to do with their own land? (This involves a key constitutional issue. You can find out more about it by searching the World Wide Web combining the keywords "constitution," "takings," and "clause.")

RECENT DEVELOPMENTS

On June 28, 2000, the Chesapeake Bay Program adopted a new Bay agreement, Chesapeake 2000: A Watershed Partnership. The new Chesapeake 2000 agreement will direct the Bay Program remediation efforts from 2000 to 2010.

Accompanying the agreement, the Chesapeake Bay Program published a status report on remediation efforts. Here is a summary of that status report.[8]

- Bay grasses covered over 68,000 acres of bottom, about 10% of their original distribution. This represented a doubling of area since 1991.
- Almost 500 miles of riparian forest[9] had been restored between 1996 and the end of 1999. The goal is 2000 miles by 2010. Eighty thousand miles of stream bank would have to be restored to return the watershed to its pre-European condition.
- The N load declined 42 million pounds per year between 1985 and 1998.
- The P load declined 6 million pounds per year.
- The number of boat waste pump-out facilities increased to nearly 600 by the end of 1999 in Virginia and Maryland. Human waste discharged from boats is an important source of N and P to the Bay.
- The Bernie Fowler "sneaker index" rose to nearly four feet. (Bernie Fowler was a Maryland legislator revered for his work toward Bay restoration. Each year Fowler waded into the Bay at the same place and measured the depth at which he could still see his sneakers. In the 1950s he could see them in more than 5' of water, but by 1989 the sneaker index was below one foot! Water clarity is one important indicator of water quality.)
- Between 1988 and 1999, 1032 miles of tributary streams were reopened to migratory fish passage. Recall that there were more than 2000 dams and impoundments on the Bay as of the early 1980s. These dams block migratory fish from reaching their historic spawning grounds.

FUTURE CHALLENGES

The Chesapeake Bay Program identified numerous challenges to Bay protection that must be addressed in the decades ahead. Among them were

- Invasive species, brought into the Bay in the ballast water of ocean-going cargo vessels (see Issue 20).
- Recovery of the Bay's oyster populations, which had virtually disappeared by 1991.
- Protection of forest and agricultural land from sprawl development (see Issues 5 and 6).
- Further removal of N and P from sewage.
- Opening up thousands of miles of tributaries to fish passage by breaching dams or building fish ladders.
- Protecting the blue crab, a symbol of Chesapeake Bay but threatened by overharvesting and pollution.
- Dealing with material dredged from harbors (Baltimore especially) and shipping channels, which may be contaminated with toxic materials, but which, even if uncontaminated, could smother bottom-dwelling organisms.

---

[8]For details go to http://www.chesapeakebay.net/pubs/snapc2k.pdf.
[9]Riparian forests are forested areas adjacent to bodies of water such as streams, lakes, etc.

Here is a summary of the Bay's status from the president of the Chesapeake Bay Foundation, William C. Baker. "Despite progress on many fronts, the bay's health has stalled at a dangerously low level."[10]

FOR FURTHER STUDY

- Go to http://www.chesapeakebay.net/c2k.htm, and http://www.cbf.org/ and assess the status of Bay remediation for yourself. Also see www.chesbay.org to assess the impact of acid rain, much of it from outside the watershed, on the Bay ecosystem.

- Research the mixing of freshwater and saltwater in estuaries and in Chesapeake Bay by referring to Thurman and Burton[11] and other Chesapeake Bay websites.[12] How efficiently does the freshwater mix with the saltwater in Chesapeake Bay? Does the sea water "flush" pollutants out of the Bay? If so, where do they go?

- In addition to sewage, sources of phosphorus for Chesapeake Bay include sediment runoff and agricultural runoff and animal waste from vast chicken and pig farms. Research and discuss the following questions.
  1. How does sediment contribute phosphorus?
  2. What is the effect of construction activity on sediment runoff and thus on phosphorus?
  3. Why is agricultural runoff other than manure a major source of phosphorus?
  4. How might phosphorus from sediment, and agricultural runoff, be minimized?
  5. What can be done about animal waste?

- Scientists studying satellite imagery have concluded that a significant amount of phosphorus was being introduced into Chesapeake Bay from sewage treatment plants in Pennsylvania, New Jersey, and New York. Study marine circulation in the coastal North Atlantic (see Thurman and Burton[13] and form a hypothesis to explain how this could happen.

- Discuss the general impact of population growth (leading to increased sewage) on the health of Chesapeake Bay. Refer back to Issues 1 to 6 for more information.

- There are, of course, other threatened bodies of water in the United States. In the west, restoration of San Francisco Bay is the subject of a concerted effort on the part of the state of California, the U.S. Bureau of Reclamation, agricultural interests in the Central Valley, and a number of public interest and environmental groups. Research the issues involved. Include in your study how water is diverted from the river systems that feed the Bay in order to grow crops in the arid Central Valley and to water lawns in California cities. The *San Francisco Chronicle* has published numerous articles on this subject, and it has also been the subject of hearings by the U.S. Congress. Lake Tahoe, one of the most beautiful lakes in North America, is gravely threatened by development, which is contributing to destructive nutrient enrichment in the lake. In 1997 the federal government joined with California and Nevada to formulate a plan to rescue the lake. Research this issue. How has development and population growth contributed to the degradation of Lake Tahoe? What remedies are proposed? How do you assess their likelihood of success?

---

[10]http://www.cbf.org/.
[11]Thurman, H.V., & E.A. Burton. 2000. *Introductory Oceanography* (Upper Saddle River, NJ: Prentice Hall).
[12]These include: Chesapeake Bay Alliance (http://www.gmu.edu./bios/Bay/acb), Chesapeake Bay Commission (http://www2.ari.net/cbc/cbc.html), Chesapeake Bay Foundation (http://www.savethebay.cbf.org), Chesapeake Bay Program (http://www.chesapeakebay.net/bayprogram), Maryland Geological Survey (http://www.mgs.dnr.md.gov), University of Maryland Center for Environmental Science (http://www.umces.edu), U.S. Environmental Protection Agency Region 3 (http://www.epa.gov/region03), U.S. National Oceanic and Atmospheric Administration (NOAA) Chesapeake Bay Office (http://www.nmfs.gov/irf/irf.html).
[13]Thurman, H.V., & E.A. Burton. 2000. *op.cit.*

ILLEGAL IMMIGRATION: BALLAST WATER AND EXOTIC SPECIES

KEY QUESTIONS

- What are exotic species?
- How are they able to cross oceans?
- What enables them to colonize foreign environments?
- Do they represent a threat to coastal ecosystems?
- What can policy makers do to protect water bodies from non-native species?

In September 1999, U.S. President Bill Clinton issued an executive order that directed the Departments of Agriculture, Interior, and Commerce, the Environmental Protection Agency, and the U.S. Coast Guard to develop an alien species management plan to blunt the economic, ecological, and health impacts of invasive species.

Agriculture Secretary Dan Glickman promised "a unified, all-out battle against unwanted plants and animal pests."

Environmentalists, meanwhile, complained that the Clinton Administration had been slow in regulating ballast discharges from freighters—one of the major pathways for exotic aquatic organisms such as the Chinese mitten crab (annual economic cost unknown)[1]; green crab (annual economic cost $44 million); and Asian clam (annual economic cost, $1 billion), which threaten native marine life in San Francisco Bay and as far north as Washington state. Another such animal is the veined rapa whelk (Figure 20-1), which has been recently discovered in Chesapeake Bay.

Since the issue involves interstate and international commerce, individual states and counties cannot, under the U.S. Constitution, regulate ballast water. The West Coast invaders, says Linda Sheehan of the Center for Marine Conservation in San Francisco, are driving out native crabs and clams and threatening local oysters—even burrowing into and weakening flood control levees, which could potentially result in huge losses from property damage during floods.

---

[1]The U.S. Bureau of Reclamation operates a series of giant pumps at Tracy, California on the delta of the Sacramento River to ship Bay water to Southern California. During 1998, they found so many mitten crabs clogging fish screens (which keep fish out of the pumps and so keep them from being cut to pieces) that the agency spent $400,000 to build a series of "crab screens" that catch the crabs before they clog the fish screens. The devices then fling them onto a conveyor belt for removal by a firm that pays for the privilege. The firm uses the mitten crabs for bait. Ironically, Chinese mitten crabs are a treasured delicacy in Hong Kong, and some entrepreneurs have explored shipping the California crabs to Hong Kong. But the state refuses to allow this, fearing that it will encourage more importation of the crabs and other exotic species, with more unforeseen consequences! (*San Jose Mercury News*. Crab migration drops off, by N. Vogel. Oct 14, 1999.)

FIGURE 20-1 The veined rapa whelk. As of June 2001, 1300 confirmed observations of this species had been recorded. U.S. East Coast estuaries have favorable temperatures and ample prey (bivalves) for the rapa whelk. (© 2001. Juliana M. Harding, Molluscan Ecology Program, Virginia Institute of Marine Science Gloucester Point, Virginia 23062)

A coalition of environmental groups and the Association of California Water Agencies asked the EPA to regulate freighter ballast water discharges under the Clean Water Act.

BACKGROUND

Silently, almost imperceptibly, the planet's oceans, seas, estuaries, and lakes are being invaded by plants, animals, bacteria, and even viruses from distant climes. These organisms are called alien, exotic, or invasive species. Sometimes their impact is negligible, rarely is it beneficial, and often it borders on the disastrous. At a January 1999 meeting of the American Association for the Advancement of Science, Cornell University ecologist David Pimentel estimated the total cost of invasive species at $123 billion a year.

Instead of remaining in an ecosystem in which all members have evolved and interacted over time, in the relative blink of an eye invasive species may be transported beyond their natural range into the presence of other organisms with which they will immediately begin to interact, and perhaps compete.

Once thrust into a new environment, an organism faces a whole new set of conditions. To survive, all living organisms must live long enough to bear offspring and thus ensure the future of their gene pool. The "aim" of exotic species is not to take over an estuary or clog a factory's water pipes, but rather to simply survive and reproduce.

Scientists believe that most nonnative organisms fail to survive in their new environment long enough to become established. And that's a very good thing. But occasionally the introduced organism finds its new home completely livable, sometimes even ideal.

Successful invasive species usually share a similar set of characteristics, according to the U.S. Coast Guard:

■ They are *hardy,* indicated by their surviving a trip inside a ship for perhaps thousands of miles.
■ They are *aggressive,* with the capacity to outcompete native species.
■ They are *prolific breeders* and can take quick advantage of any new opportunity.
■ They *disperse rapidly.*

Rapid dispersal is facilitated by having a planktonic larval stage, which allows the juveniles to be carried far and wide by currents. Such an introduced species often spreads rapidly, especially when predators and pathogens normally encountered in its home range are absent from the new environment, or when they are better able to feed than their new neighbors. (Or if they find their new neighbors especially tasty!)

In the above scenario, alien species flourish and potentially can reach astonishingly high population levels. Often, native species are displaced, or "outcompeted" by the invaders. Then the situation is often called an *invasion.*

Invasive species can inflict damage on ecosystems by:

■ Outcompeting native species
■ Introducing parasites and/or diseases
■ Preying on native species
■ Dramatically altering habitat[2]

Ballast Water

Most invasive species are brought to new shores in the *ballast water* of ships (Figure 20-2), but animals dumped into an estuary from aquariums or accidental releases from aquaculture facilities may also contribute.

What is ballast water? Ballast water is carried by ships in special tanks to provide stability, optimal steering, and efficient propulsion. According to the U.S. Coast Guard, the use of ballast water varies among vessel types, among port systems, and with cargo and sea conditions.

How much ballast water is involved? The National Oceanographic and Atmospheric Administration (NOAA) has calculated that 40,000 gallons (150,000 liters) of foreign ballast water are dumped into U.S. harbors each minute.

The problem with ballast water is very simply stated: Ballast water is taken up by a ship in ports and other coastal regions, in which the waters may be usually rich in planktonic (small, floating, or weakly swimming) organisms. It may be released at sea, in a lake or a river, or in the open ocean along coastlines—wherever the ship reaches a new port. As a result, a myriad of organisms is transported around the world within the ballast water of ships and is released. Here are two examples.

Scientists studying an Oregon bay counted 367 types of organisms released from ballast water of ships arriving from Japan over a four-hour period[3]! Another study documented a total of 103 aquatic species introduced to or within the United States by ballast water and/or other mechanisms, including 74 foreign species.[4]

---

[2]www.wsg.washington.edu/pubs/bioinvasions/bioinvasions5.html.
[3]www.darwin.bio.uci.edu/~sustain/bio65/lec09/b65lec09.htm.
[4]www.ucc.uconn.edu/~wwwsgo/ballast.html.

FIGURE 20-2 Ship dumps ballast water after entering harbor. (Northeast Sea Grant/ photo by L. David Smith)

INVASIVE SPECIES AND CHESAPEAKE BAY

There is growing concern about invasive species' impact on Chesapeake Bay (see Issue 19). How much of a potential problem can be assessed by these 1995 statistics from the Chesapeake Bay Commission:

- More than 90% of vessels arriving at Chesapeake Bay ports carried live organisms in ballast water, including, but not limited to, barnacles, clams, mussels, copepods, diatoms, and juvenile fish.
- Nonindigenous species have been responsible for paralytic shellfish poisoning, declining commercial and sport fisheries, and even cholera outbreaks!
- The two ports of Baltimore and Norfolk alone receive 2,834,000, and 9,325,000 metric tons of ballast water, respectively, each year, and this water originates from nearly fifty different foreign ports.

A database of organisms nonindigenous to Chesapeake Bay, prepared by the Marine Invasions Research Lab of the Smithsonian Environmental Research Center,[5] lists 160 species and classifies another 42 as of uncertain origin.

Today, ballast water appears to be the most important means by which marine species are transferred throughout the world. As you saw above, the transfer of organisms in ballast water has resulted in the unintentional introduction of hundreds of freshwater and marine species to the United States and elsewhere.

---

[5]http://160.111.110.108/species.htm.

Furthermore, the rate of new invasions from ballast water has increased in recent years.[6] To the extent these unwelcome visitors do economic damage, they make up a generally hidden cost of world trade.

INVASIVE SPECIES ALONG THE PACIFIC COAST OF THE UNITED STATES

Chinese mitten crabs, another invasive species affecting San Francisco Bay and the Sacramento River Delta, are described by some scientists as "burrowing fiends," digging burrows that can significantly weaken levees (embankments built to prevent flooding) in a region that is prone to dangerous floods.

Question 1: If the costs incurred during a flood are in part due to ships involved in international trade, how can these costs be fairly apportioned? Is it fair for only those people who are affected by floods in California to pay for the hidden costs incurred as a result of invasive species? Is it reasonable to price imported goods cheaply and then expect local residents to bear the cost of flooding resulting from this trade? Suggest some possible solutions to this problem, but remember, localities and states do not have the right under the Constitution to regulate international trade.

Another growing problem is posed by the European green crab—a recent import that is affecting coastal California and seems to be working its way up the coast to Oregon and Washington. This aggressive predator prefers clams to oysters but could prey on baby Dungeness crabs (an economically important species) and smaller shore crabs.

And Washington State's oyster farmers are already uneasily coexisting with another exotic: the oyster drill that came from Japan with Pacific oysters. Already Washington oyster growers have had to abandon habitat overrun by the oyster drill.

By now you should have seen ways that invasive species can materially affect the U.S. economy as well as our environment.

THE VEINED RAPA WHELK (*RAPANA VENOSA*)

Rapana venosa (Figure 20-1) is a predatory gastropod. Juliana Harding and Roger Mann, two researchers at the Virginia Institute of Marine Sciences (VIMS), are studying the whelk's impact on Chesapeake Bay.[7] The following is a summary of their research.

Discovery of *R. venosa* was purely by accident: A routine trawl in the lower reaches of Chesapeake Bay turned up an unknown organism, which was ultimately identified by scientists at the Smithsonian Institution in Washington DC and by a Russian biologist at the Moscow Academy of Sciences as *R. venosa. R. venosa* has left a trail of destruction behind itself in its wanderings: Most recently, an oyster population in the Black Sea had been decimated by *R. venosa* predation.

---

[6]www.serc.si.edu/invasions/ballast.htm#Box 1.
[7]Their progress can be monitored at www.vims.edu/fish/oyreef/rapven.html. They also publish an on-line newsletter at www.vims.edu/fish/oyreef/rud.html.

Excited, the researchers conducted another trawl, which turned up no individuals, but did collect two live masses of *R. venosa* eggs, which they returned to the lab and set about to hatch, in a script beginning to take on similarities to the film "**Aliens**"!

As the eggs began to hatch, the scientists were eager to determine the tolerance of the hatchlings to variations in temperature and water salinity.

Question 2: Why do you think they were interested in these data?

The scope of potential contamination of the Bay by *R. venosa*, which is native to the Sea of Japan, and other introduced species can be appreciated by an estimate of the volume of ballast water dumped into Chesapeake Bay ports during 1998 solely by ships from ports with active *R. venosa* populations: 15 million tons! And, since the entrance to Chesapeake Bay is the site of considerable coastal shipping, infestations of harbors from Boston to Charleston, South Carolina was a distinct possibility.

Subsequently, Harding and Mann learned something about the whelks' habitat preferences and diet. The whelks preferred hard sandy bottom into which they would quickly burrow: A 6" whelk could completely hide itself in less than an hour, leaving only its purplish colored siphon exposed, which it would instantly withdraw if disturbed. *R. venosa* spends at least 95% of its life burrowed, but it can and does move while burrowed, at speeds up to one body length per minute. They learned that the whelk can feed and mate while completely buried.

Question 3: Considering the whelk's preferred habitat, propose how scientists could determine its potential range in an estuary like Chesapeake Bay.

R. venosa's preferred diet was hard clams, but it would eat oysters, soft clams, or mussels if its favorite food was unavailable. Unfortunately for the clams, they share the whelk's habitat. And finally, there is a "healthy" hard clam commercial fishery in Chesapeake Bay.

The researchers were also interested in the whelk's predators if any, in *R. venosa*'s home waters. There are few: Octopi eat the whelks in the waters of southern Russia and the Black Sea, but there are none in Chesapeake Bay. Other native whelks in Chesapeake Bay prey on smaller individuals of *R. venosa*, but there is an interesting twist: *R. venosa*'s shell is much thicker and more robust than native whelks; moreover, *R. venosa*'s boxy shape means the creature is hard for other whelks to eat as adults. So if the whelks can survive to adulthood, they have little to fear from the natives in Chesapeake Bay.

The researchers next turned their attention to what could prove to *be R. venosa*'s "Achilles heel," the egg and juvenile stages. They concluded that as eggs the whelks may be most vulnerable, since migrating fish could eat the bright yellow egg cases, or dislodge them, causing damage and perhaps death to the developing eggs.

In summary, little was known about the actual distribution and impact of *R. venosa* in its new home, Chesapeake Bay, the most economically productive estuary in the United States. But researchers are taking the potential for serious impact very seriously.

Question 4: What authority do state governments and localities have in controlling the means by which the whelks are spread?

THE ZEBRA MUSSEL

Some invasive species belong to the Phylum Mollusca, which includes clams, oysters, and snails, among others. Mollusks can be found in marine, brackish, and fresh water. The most notable invasive mussel introduction so far is the case of the zebra mussel *Dreissena polymorpha* (Figure 20-3), a native of eastern Europe. (The original description of this species, from 1769, was of populations in the Ural River and Caspian Sea of the former Soviet Union.)

The zebra mussel has caused serious economic and ecosystem impacts, with costs projected in 1998 to be $5 billion over the next ten (1998–2008) years, absent controls.[8] But a similar estimate published in 1993 forecast $3 billion damage for the forthcoming decade (1993–2003), so you can see that the problem is getting worse. And a team of ecologists at Cornell University estimated the cost of zebra mussels to be $3 billion a year!

Zebra mussels can destroy entire colonies of native mussels by interfering with such basic functions as respiration, reproduction, feeding, growth, and movement.

One of the reasons for their success is their proficiency at breeding. Up to one million eggs can be laid by one female in a spawning season. Upon hatching, the larvae may be dispersed by currents after which as juveniles they settle to the bottom and attach. Importantly, they have trouble keeping attached in water velocities above around two meters per second.

Moreover, zebra mussels are filter feeders, which means they have specialized organs for filtering food, mainly algae, out of the water.

Question 5: Suggest a way in which the introduction of zebra mussels into a water body contaminated with too many planktonic algae might actually be beneficial.

---

[8]www.glc.org/ans/96rpt.html.

FIGURE 20-3 Zebra mussels cover a crayfish. This invader has spread to 20 U.S. states and can tolerate estuarine water. (GLSGN Exotic Species Library)

In the Great Lakes, zebra muscle concentrations of up to 700,000 per square meter have been reported.

Question 6:   At this density, how many zebra mussels were there per square centimeter?

Question 7:   At this density, how much water could have been filtered clean of algae by one square meter of zebra mussels each day?

Question 8: At this density, how many zebra mussels would be needed to filter a cubic kilometer of algal-polluted lake water each day? This would be equal to a lake with an area of 100 square kilometers and an average depth of ten meters.

Question 9: Finally, assess the potential adverse impact of introducing zebra mussels into a polluted water body to clean it of algae.

Zebra mussels were first discovered in North America in 1988, when a few were found in Lake St. Clair, Michigan. The initial introduction is believed to have occurred in 1985 or 1986 via ballast water, perhaps in the holds of cargo ships sent to pick up iron ore or grain. Since then, zebra mussels have proven to be a very costly pest to municipal and industrial water users.

The impact on industries drawing water from the Great Lakes was rapid and caused shutdowns due to severe flow reductions as mussels attached to intake structures and the insides of pipelines.

Utilities that operate power plants relying on lake water for cooling have been among the most heavily impacted. Since 1989, power plants, water utilities, industrial facilities, and navigation lock and dam operators have spent more than $70 million trying to control and manage zebra mussel infestations.[9]

The spread of zebra mussels continues: By 1999 zebra mussels were found by the U.S. Geological Survey in the Missouri River.

It is clear that without a massive effort (and perhaps even with one) the spread of zebra mussels will not be contained.

FOR FURTHER STUDY

- Go to the U.S. government websites for Commerce, Agriculture, EPA, Interior, and the Coast Guard to assess those agencies' progress towards developing the Alien Species Management Plan as directed by former President Clinton.
- Design a research project to study and measure the evolving impact of *R. venosa* on Chesapeake Bay, as well as its potential impact on other harbors along the East Coast. Suggest means by which the whelk's impact might be reduced or controlled.

---

[9]www.wsg.washington.edu/pubs/bioinvasions/bioinvasions5.html.

CATCH OF THE DAY: THE STATE OF GLOBAL FISHERIES

KEY QUESTIONS

- What is the state of global fisheries?
- What is bycatch?
- What is the environmental impact of commercial fishing?
- How much seafood do we eat?
- How important is seafood as a protein source?
- Is "sustainable fisheries" an oxymoron?

Wild Oats Markets, a nationwide chain of more than 75 grocery stores, issued a press release[1] stating that it would no longer sell North Atlantic swordfish, marlin, orange roughy, or Chilean sea bass because these species are endangered due to overfishing.

Paul Gingerich, Meat and Seafood Purchasing Director, commented, "Floods, drought and overgrazing that affect food sources on land can be readily seen and measured. The effects of overfishing cannot be easily seen in the oceans. We need to be proactive to save these species for future generations."

Will the other 246,000[2] U.S. grocery stores follow suit? Are the species in question, and others, really endangered?

THE IMPACT OF GLOBAL FISHERIES

Even though less than 1% of global caloric intake comes from fish,[3] the importance of fisheries to the global and many national economies cannot be overstated. Consider the following information on fisheries worldwide[4]:

- The value of the international fish trade for 1994 was $47 billion.
- The combined value of canned, fresh, and frozen fishery products in the United States in 1996 was over $2.9 billion.
- Nearly 85,000 people were employed in processing and wholesale jobs alone in the United States in 1995.

---

[1]Available at www.wildoats.com.
[2]*Statistical Abstract of the U.S. 1996–97,* Table No. 1283, p. 770.
[3]Pimentel, D., & M. Pimentel (eds.). 1996. *Food, Energy, and Society.* (Niwat, CO: University Press of Colorado).
[4]This information was obtained from a variety of sources, including the National Marine Fisheries Service, *Statistical Abstract of the U.S. 1998,* and World Resources 1996–97.

- The economies of many countries such as Iceland, Peru, and Norway depend heavily on fish product exports.
- In eastern Canada, the closure of the cod fishery cost at least 40,000 jobs, in a country with a population one-tenth that of the United States.
- Of the $752 the average American spent on meat in 1995, $97 (13%) was for seafood.

Although the contribution of fish to the diet of humans may seem small if we simply count calories, it becomes critical if we consider protein: 16% of global animal protein is provided by fish, while in the Far East, where most humans live, nearly 28% of animal protein comes from fish.[5] In developing countries worldwide, where population growth rates are ominously high, 950 million people depend on fish as their primary source of protein.[6]

In addition to serving as a basic food source, fish is increasingly considered by affluent westerners to be a "health food." Fatty fish like salmon and mackerel have relatively high levels of omega-3 fatty acids, which have been shown in clinical studies to reduce the risk of heart attack by 50 to 70%.

Finally, fish and fish by-products, representing as much as one-third of wild-caught fish, are a mainstay of the pet food industry and are used as a constituent of animal feed as well.

ENVIRONMENTAL COSTS OF FISHING

Commercial fishing can be a very expensive as well as an environmentally costly activity. First, many fishing methods destroy habitat. Consider trawling, which is typically done for shrimp and other bottom-associated species like Atlantic cod. In this method, a 10 to 130 m (33–426 ft) long net is scraped across large areas of the bottom, collecting virtually everything in its path, including endangered sea turtles. Trawling, which became popular with the advent of the diesel engine in the 1920s, is practiced worldwide on virtually every different bottom type.[7] Saturation trawling, in which the net is repeatedly fished in an area until virtually no fish or shrimp are left, has been compared to clearcutting a forest. The comparison is appropriate: Trawling heavily damages sessile benthic organisms like sponges, hydroids, and tube-dwelling worms and displaces associated fauna like fish and crustaceans. Complete recovery in both cases may take centuries. The comparison between trawling and forest clearcutting breaks down when one considers the area involved annually. Approximately 100,000 km² (38,000 mi²) of forest are lost annually, whereas an area 150 times as large is trawled.[8]

Innovations like TEDs (Turtle Excluder Devices) shunt large objects like sea turtles out of the net. The use of TEDs, however, is not universal and is not entirely successful either.

Extremely destructive methods like dynamiting and poisoning still are used in some areas.

Second, commercial overfishing threatens fish stocks, which are under stress from coastal environmental degradation due to overdevelopment and industrial, municipal, and agricultural pollution. A 1994 United Nations Food and Agricultural Organization (FAO) analysis of marine fish resources concluded that 35% of 200 top marine fisheries

---

[5]Food and Agricultural Organization of the United Nations. 1993. Marine fisheries and the law of the sea: A decade of change. FAO Fisheries Circular No. 853. Rome.

[6]World Resources Institute. 1996. *World Resources 1996–97: A Guide to the Global Environment.* (New York: Oxford University Press).

[7]Waitling, L., & E.A. Norse. Disturbance of the seabed by mobile fishing gear: A comparison to forest clearcutting. *Conservation Biology, 12:* 1180–97.

[8]Ibid.

FIGURE 21-1 Bycatch includes sea turtles, mammals, birds, invertebrates, and fish, like this oceanic sunfish caught in a Japanese driftnet in the Tasman Sea. (Photo courtesy of Greenpeace)

were overexploited (i.e., yielded declining landings); 25% were mature (and thus on the verge of endangerment if stressed); and 40% (largely in the Indian Ocean) were still developing.

Significantly, *no* major fisheries, according to the study, were undeveloped.[9]

Third, at least 25% of the commercial catch is unused. This quantity is known as *bycatch* and refers to undersized, low-value, and nontarget species (fish, crabs, etc.). Bycatch is often returned to the water dead, or dies soon after (Figure 21-1). The fishing activity near the top in bycatch is trawling for shrimp. In addition to damaging the ocean bottom, as much as 90% of the trawl contents may be nontarget and hence unused species, sometimes called "trash fish" by fishers.

Finally, the environmental cost of commercial fishing extends to the pollution associated with the manufacture, transportation, and use of equipment (like fishing boats) and supplies; fuel spills; and transportation and refrigeration of fishery products.

CASE STUDY: THE SLIMEHEAD AND PATAGONIAN TOOTHFISH

Would you eat a fish called a slimehead? Or a Patagonian toothfish? Probably not, so clever marketing specialists transformed these into popular items by renaming them as "orange roughy" (Figure 21-2a) and "Chilean sea bass" (Figure 21-2b). The former was popular in the mid-to late 80s, whereas the latter's popularity is just peaking.

Unfortunately, renaming the fish changed neither their biology nor their fate. The orange roughy is a classic example of underestimating the importance of gaining a complete understanding of a species' biology before exploiting it as a fishery. The Chilean sea bass is yet another reminder that we refuse to learn from our mistakes. Both cases demonstrate the power and potential environmental destructiveness of effective marketing.

---

[9]Food and Agricultural Organization of the United Nations. 1996. Chronicles of marine fisheries landings (1950–1994): Trend analysis and fisheries potential. FAO Fisheries Technical Paper No. 359, Rome.

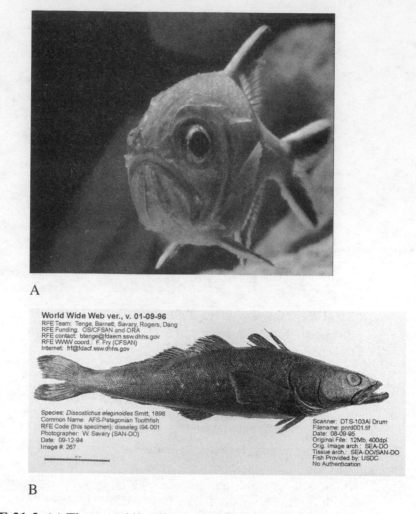

A

B

FIGURE 21-2 (a) The erstwhile *slimehead,* successfully marketed as the orange roughy. (Jean-Paul Ferrero/Jacana, Photo Researchers, Inc.) (b) The fish formerly known as the Patagonian toothfish, now available in Western restaurants and seafood shops as the Chilean sea bass. It is not at all closely related to the group of fishes commonly known as sea basses. (W. Savary, Center for Food Safety and Applied Nutrition)

The orange roughy fishery began off New Zealand in 1978 and quickly exceeded 35,000 tons[10] (31.8×10^6 kg). Unfortunately, because the species is long-lived (how ironic!), reaches sexual maturity late in life, and doesn't produce profuse numbers of offspring, by 1990 the harvest had been reduced over 70%. Maximum sustainable yield, the amount fisheries managers estimate can be annually harvested without damaging the population, had been estimated to be 7,500 tons (6.8×10^6 kg), which is still lower than the 1990 harvest.

Chilean sea bass have a similar biology and thus are likewise vulnerable to overfishing by powerful modern fishing methods.

WORLD FISH CATCH

Information on the total annual world fish catch is compiled by the Food and Agricultural Organization (FAO) of the United Nations. Table 21-1 below contains this information (not including aquaculture) for the years 1950 to 1997.

---

[10]McGinn, A.P. 1999. Safeguarding the health of the oceans. Worldwatch Paper 145, Worldwatch Institute.

Table 21-1 ■ World Fish Catch, 1950 to 1997 (Modified from *Vital Signs, 1999*)

Year	Total Catch (Million tonnes*)
1950	19
1955	26
1960	36
1965	49
1970	58
1975	62
1980	67
1985	79
1990	86
1991	85
1992	86
1993	87
1994	93
1995	93
1996	95
1997	94

*Note that a "tonne" is the same thing as a "metric ton" and equals 1000 kg (2200 lb).

Question 1: On the axes below, make a graph of world fish catch (in million tonnes) for the period from 1950 to 1997. (You can do this manually or use a spreadsheet such as Excel.)

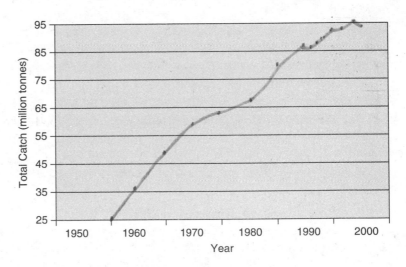

Question 2: Interpret the graph: What trends can you infer?

The trend that I can infer is each year more fish is caught then the last year.

Question 3: Now, we want to calculate the average *annual* rate at which total catch changed for a specified period. Do this now for the 10-year periods 1950–60; 1960–70; 1970–80; 1980–90; 1990–97. Use the compound growth equation (**$k = (1/t)\ln(N/N_0)$**), where k = the percentage increase per unit time; t = the number of years; N = total catch at the end of the given time; and N_0 = total catch at the beginning of the period (see *Using Math in Environmental Issues*, pages 15–16).

Question 4: The late economist, Julian Simon, in his 1996 book *Ultimate Resource 2*[11] (p. 104) stated: "No limit to the harvest of wild varieties of seafood is in sight. . . . It is still rising rapidly." Is this statement consistent with the trends you just observed? Why or why not? How would you project the future of capture (i.e., nonaquaculture) fisheries?

Table 21-2 below shows world population and total fisheries catch.

---

[11]Simon, J.H. 1998. Ultimate Resource 2. Priceton University Press, Princeton, N.J.

Table 21-2 ■ World Population and Total Fisheries Catch, 1950–1997

Year	Total Catch (Million tonnes)	World Population (billions)
1950	19	2.556
1955	26	2.780
1960	36	3.039
1965	49	3.345
1970	58	3.707
1975	62	4.086
1980	67	4.454
1985	79	4.851
1990	86	5.277
1991	85	5.359
1992	86	5.442
1993	87	5.523
1994	93	5.603
1995	93	5.682
1996	95	5.761
1997	94	5.840

Question 5: Use the data in Table 21-2 to calculate per capita (i.e., per person) fish consumption. Report as $kg \cdot person^{-1}$ (i.e., "kilograms per person"). Fill in the empty column below with your calculations.

Year	Per Capita Fish Consumption ($kg \, person^{-1}$)
1950	
1955	
1960	
1965	
1970	
1975	
1980	
1985	
1990	
1991	
1992	
1993	
1994	
1995	
1996	
1997	

Question 6: On the axes below, make a graph of per capita fish consumption (in kg per person) for the period from 1950 to 1997.

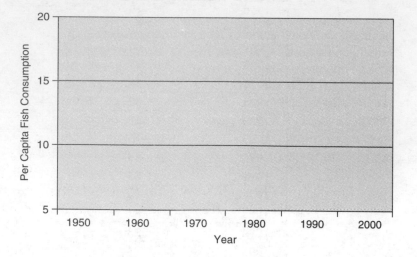

Question 7: Describe any trends in your graph.

Question 8: Calculate average annual percent change in per capita fish consumption for the periods 1950–60; 1960–70; 1970–80; 1980–90; 1990–97.

Question 9: With global population growing at about 1.24% per year, do you think that capture fisheries can supply demand on fish stocks of a burgeoning human population? Justify your answer.

Question 10: Summarize what you have learned about global fisheries and fish consumption from answering Questions 1 through 9.

Question 11: The lives of more than 1 billion people are threatened by starvation or malnutrition. In the 1960s and 1970s some thought that the solution to world hunger was protein derived from global fisheries. Why has this solution not materialized? Do you think food from capture fisheries can play a role in abating world hunger?

Question 12: Sylvia Earle, former director of the National Oceanic and Atmospheric Administration, has said "People who want to make a difference can choose not to eat fish that are more important swimming alive in the ocean than swimming in lemon slices and butter."[12] Do you agree? Why or why not?

FOR FURTHER STUDY

- Search the Web using the key words "Chilean sea bass." Did you find references on its biology, fishery, and conservation or did your "hits" contain mostly recipes?
- An outstanding game simulating world fisheries is *FISH BANKS, LTD*. Information on this game, which has been played by heads of state and other international and national policymakers may be obtained from the Institute for Policy and Social Science Research, Hood House, University of New Hampshire, 89 Main St., Durham, NH, 03824. The Institute also maintains a website for the game at http://www.unh.edu/ipssr/Lab/FishBank.html.
- Identify and quantify issues dealing with the environmental costs of commercial fishing. Consult the National Marine Fisheries Service's National Marine Fisheries

[12]National Oceanic and Atmospheric Administration Sustainable Seas Expedition website. http://www.sustainableseas.noaa.gov/aboutsse/liveevents/liveevents1999/april26chat.html.

Service website (http://www.kingfish.ssp.nmfs.gov) and others (e.g. Natural Resources Defense Council http://www.nrdc.org).

■ Shark populations are dwindling, some to dangerously low levels. Research this issue. Which sharks are threatened? What are the causes? What is being done to help sharks recover? What role do sharks play in their environment? What might be the consequences of extinction of some shark species?

■ An article on fisheries in *Outdoor Life*[13] began

> To say that we "manage" our marine fisheries is like saying that CPR is a terrific way to manage heart disease. Instead of actual management, too often the process is one of recurrent ignorance, greed and last-minute legislation. We have devastated many of our fisheries to such as extent that Dr. Ian Fletcher, chief scientist at the Great Salt Bay Experimental Laboratory . . . says, "I think it a wonder there's any fish on the market, given all the insults."

Experts assert that the future of marine fisheries is wholly dependent on effective management. Research how fisheries are managed by consulting the Natural Resources Defense Council website. Do you concur with the *Outdoor Life* article? Explain your reasoning.

■ Make a list of the seafood you have eaten in the last year, then go to the website http://www.mbayaq.org/efc/efc_oc/dngr_food_watch.asp and fill in the table with information from that website.

Seafood Item	Where's It From?	Level of Fishing	More Information

Based on your seafood chart above, analyze whether your diet was seafood friendly and was a part of sustainable fisheries or whether it contained seafood whose capture or culture had a significant environmental impact.

■ The following talk show lasts about 48 minutes. You should read the questions and take notes as you listen (http://www.wamu.org/ram/1999/r1991123.ram).

Question 1: What does the Sustainable Fisheries Act of 1996 require the National Marine Fisheries Service (NMFS) to do?

Question 2: Summarize the state of the Atlantic swordfish fishery.

Question 3: Do you think that monitoring juvenile swordfish imports using the Certificate of Origin is effective? Why or why not?

Question 4: A caller notes that U.S. consumers would never know that there was a fisheries crisis after visiting a restaurant or supermarket. Do you agree? Why do you think this is so?

Question 5: Another caller called global fisheries a "modern tragedy of the commons" and claimed that we have no problem with hogs, chickens, and pigs because they are privately owned. What is the "tragedy of the commons" (look it up if necessary)? Would partitioning the oceans into privately held sections solve the problem? Discuss.

[13]Boyle, R. H. 1995. Management school. *Outdoor Life,* April 1995: 54–59.

AQUACULTURE

KEY QUESTIONS

- What are the advantages of aquaculture?
- What are its environmental impacts?
- How fast is aquaculture growing?

Although "capture" fisheries are nearly fully exploited, and in many cases severely over-exploited, total fish production is increasing. This increase is a result of growth of *aquaculture,* mainly in China. The Chinese have been masters of aquaculture for centuries, if not millennia. Today Chinese farmers routinely integrate rice cultivation with fish farming, growing several species of carp in rice paddies. The advantages are clear: Carp may be filter feeders or herbivores; in either case, they do well in ponds that have been fertilized with manure. The manure feeds grasses, which may be fed on by herbivorous carp, and the grasses provide habitat where phytoplankton and zooplankton live, which in turn are food for filter-feeding carp. So this style of aquaculture places little stress on already threatened marine environments. By contrast, several countries (Norway, Chile, and the United States, for example) are beginning to grow carnivorous species like salmon, which actually put additional stress on wild fish stocks, since up to 5 kilograms of wild-caught fish must be fed to salmon to produce a kilogram of meat.

Two other concerns arise out of large-scale salmon farming: Waste production and the deterioration of the wild gene stock. Salmon kept in pens and fed large quantities of fish meal produce in turn large quantities of concentrated waste: In Norwegian salmon farms, for example, the fish produce as much waste as Norway's 4 million humans. Additionally, farmed salmon are bred for rapid growth and not to develop traits that will aid their survival in the wild. Salmon that escape their pens can and do breed with wild salmon, however, and could in time reduce their wild cousins' ability to survive in the open ocean.

Shrimp *mariculture* is growing rapidly as well (Figure 22-1). Tiger shrimp, the main variety grown, is a tropical crustacean; thus, areas developed for shrimp farms tend to be areas in the coastal tropical ocean, itself under severe strain from the growth of human populations. (see Issues 1-3).

The greatest problems posed by marine fish-farming relate to land-use changes.[1] Tens of thousands of hectares of salt-water wetlands are lost annually by conversion to shrimp mariculture alone. Shrimp farming is stressful in at least two ways. First, areas devoted to shrimp farming tend to be cleared mangrove swamps (Figure 22-2). Mangroves (salt-

---

[1]The weakening of genetic diversity of fish stocks poses a problem as well.

FIGURE 22-1 Aerial view of shrimp farm ponds in Honduras. (©Tim Wright/CORBIS)

tolerant trees) are of critical importance as nurseries for juvenile fish and other marine animals, and the mangrove roots trap and bind sediment washing offshore from rivers. This sediment, if left unchecked, can smother offshore reefs. Reefs, in turn, are important fishing grounds for many subsistence fishing communities, especially in Southeast Asia.

Mangroves also provide important protection to the coast from the impacts of tropical storms. Ironically, the mangroves cleared for shrimp farms are more productive biologically than the farms that replace them, in that the mangroves could supply a greater harvest of fish. (For example, the impact that the loss of the mangrove ecosystem has on local Ecuadorian fishermen and their families is illustrated by a study concluding the loss of one acre of mangroves will result in a drop in the harvest of wild shrimp and fish by 676 pounds per year.[2]

Second, shrimp are fed fish meal (prepared from wild-caught fish), which, like salmon, puts additional stresses on wild stocks.

SOME ADVANTAGES OF AQUACULTURE

Aquaculture, if carried out in a manner sensitive to land-use conservation and the avoidance of additional stresses on marine fish stocks, could be an important addition to a human food base that is becoming threatened by land degradation, pollution, and the addition of 80 million new mouths to feed each year.

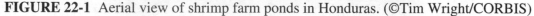

[2]www.earthsummitwatch.org.

FIGURE 22-2 Boys collect illegally cut mangrove trees at the site of a future shrimp farm, Honduras. (©Tim Wright/CORBIS)

The Worldwatch Institute's President Lester Brown is an agricultural expert. Here is what he said recently about the advantages of fish farming.[3]

> Cattle require some 7 kilograms of grain to add 1 kilogram of live weight, whereas fish can add a kilogram of live weight with less than 2 kilograms of grain. Water scarcity is also a matter of concern since it takes 1,000 tons of water to produce 1 ton of grain. But the fish farming advantage in the efficiency of grain conversion translates into a comparable advantage in water efficiency as well, even when the relatively small amount of water for fish ponds is included.

THE GROWTH OF AQUACULTURE

From 1984 to 1995, world aquaculture production rose from 6.9 to 20.9 million metric tons (6.9×10^9 to 20.9×10^9 kg), while in China it increased from 2.2 to 12.8 million metric tons (2.2×10^9 to 12.8×10^9 kg) over the same interval.[4] Fish-farming may soon overtake cattle ranching as a source of protein if these trends continue.[5]

Averaged over the period 1995–1997, world aquaculture production was 33,684,000 tonnes (this number includes harvesting of plants). For the same period, aquaculture production in Asia was 30,577,600 tonnes.

---

[3]http://www.worldwatch.org/chairman/issue/001003.html.

[4]Worldwatch Institute. 1998. *Vital Signs* (New York: W.W. Norton and Co.).

[5]www.worldwatch.org/chairman/issue/001003.html.

Question 1:   What percentage of global aquaculture production took place in Asia?

Question 2:   Using the following data, plot the growth of aquaculture and wild fish catch ("Total Catch") over the period 1950–1998.

World Fish Catch, Aquaculture, and Total Harvest, 1950–98[6]

Year	Total Catch (million tons)	Aquacultural Production (million tons)	Total Harvest (million tons)
1950	19.2	1.5	20.7
1951	21.5	1.7	23.1
1952	22.9	1.7	24.6
1953	23.7	1.8	25.5
1954	25.2	1.9	27.2
1955	26.4	2.1	28.4
1956	27.8	2.1	29.9
1957	28.6	2.3	30.9
1958	30.2	2.5	32.7
1959	33.4	2.8	36.2
1960	36.4	3.0	39.4
1961	39.5	3.2	42.7
1962	42.9	3.3	46.2
1963	44.2	3.4	47.6
1964	48.5	3.6	52.0
1965	49.0	3.7	52.7
1966	52.6	3.9	56.5
1967	55.6	4.0	59.6
1968	56.5	3.9	60.4
1969	57.4	3.7	61.1
1970	58.2	3.6	61.8
1971	62.4	3.8	66.2
1972	58.4	3.8	62.2
1973	59.0	3.9	62.8
1974	62.6	4.0	66.6
1975	62.4	4.1	66.5
1976	64.6	4.8	69.4
1977	63.4	4.8	68.2
1978	65.3	4.8	70.1
1979	66.1	5.0	71.1
1980	67.0	5.2	72.1

[6]Compiled by Worldwatch Institute (http://www.worldwatch.org) from U.N. Food and Agriculture Organization Yearbooks.

Year	Total Catch (million tons)	Aquacultural Production (million tons)	Total Harvest (million tons)
1981	69.4	5.4	74.8
1982	71.1	5.6	76.7
1983	71.6	5.9	77.5
1984	77.6	6.7	84.3
1985	79.1	7.7	86.8
1986	84.6	8.8	93.4
1987	85.0	10.1	95.1
1988	88.6	11.7	100.3
1989	89.3	12.3	101.6
1990	85.9	13.1	99.0
1991	84.0	13.7	97.7
1992	85.0	15.4	100.4
1993	86.0	17.8	103.8
1994	91.0	20.8	111.8
1995	92.0	24.4	116.4
1996	93.0	26.8	119.8
1997	93.0	28.8	121.8
1998	86.0	30.7	116.7

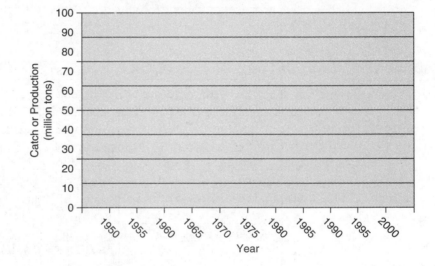

Question 3: Describe any trends apparent from your graphs, for example, which source is growing or changing fastest? Have there been intervals when rates of change differed from others? List them.

"Finfish" such as tilapia, salmon, carp (several species), and flounder comprise about one-half of aquaculture production. Mollusks, mainly oysters and mussels, account for one-fourth. Shrimp and prawns (i.e., crustaceans) make up the final fourth. Two-thirds of aquaculture takes place in inland rivers, lakes, ponds, and artificial tanks. Coastal marine aquaculture (mariculture), which includes the margins of bays, in bays, and the open ocean, accounts for the remainder.

Aquacultural output, like capture-fisheries production, is used for both fish meal and canned, frozen, and fresh products. The growth rates for aquacultural production are remarkable. Shrimp mariculture, alone grew by 350% between 1984 and 1993, in response to market demands and often to government subsidies.[7]

South America amply illustrates these trends. Below are data (in metric tonnes) for Chile and Ecuador.

Country	Total production (1996 av)	Diadromous fish	Mollusks, crustaceans
Chile	215,779	196,144	19,472
Ecuador	116,800		116,218

In the case of Chile, the diadromous[8] fish are almost entirely salmon, and in Ecuador's case, the product was shrimp. Shrimp and salmon culture impose significant environmental costs, similar in ways to the impact of terrestrial meat production, as we have seen above.

Many countries actively subsidize mariculture. But subsidies and a growing market do not always produce a winning combination, as the following case study from Ecuador will illustrate.

FOCUS: SHRIMP MARICULTURE IN ECUADOR

Shrimp farming is an important form of income for tropical coastal countries.[9] Ecuador's shrimp exports, about 60% of which come from ponds, was the second largest foreign exchange earner after oil in the early 1990s. Shrimp ponds are often constructed by cutting down mangrove habitats (Table 22-1) along tropical coastlines. This activity has been responsible for the loss of two-thirds of mangrove forests in the Philippines, more than half of Thailand's mangroves, and more than a fourth of Ecuador's.[10]

Read the summary below[11] from Earth Summit Watch, and answer the questions that follow.

Ecuador has traditionally been the largest supplier of shrimp to the United States; however, in recent years Ecuador's maricultural shrimp industry has suffered losses. This can be partially explained by market forces. Fluctuations in prices and increased competition from Asia, which uses more efficient methods, have both hurt Ecuador's exports. However, the most important factor is the current scarcity of post-larval (PL) shrimp caused by the destruction of the mangrove habitat.

Shrimp mariculture requires the creation of ponds below the high tide line, which is often accomplished by clearing the mangrove forests. These ponds are then stocked with PL shrimp (*Pennaeus vannamei* have proved to be the most productive species in Ecuador), which are harvested upon maturation with nets. The PL shrimp are supplied by

---

[7]Malcolm C., M. Beveridge, L.G. Ross, & L.A. Kelly. 1994. Aquaculture and biodiversity. *Ambio,* December 1994.

[8]Diadramous fish are those that live most of their lives in saltwater but enter freshwater to spawn.

[9]http://darwin.bio.uci.edu/~sustain/shrimpecos/louis.html.

[10]www.seaweb.org/background/book/aquaculture.html.

[11]The full story is available at www.colby.edu/personal/t/thtieten/aqua-ecua.html.

Table 22-1 ■ Ecuador: Evolution of Mangroves, Shrimp Ponds, and Saltlands Areas, 1969–1996 (In Hectares)[12]

	1969	1991	1995
Area mangroves	203,624.00	162,186.55	149,570.05
Shrimp ponds	0.00	145,998.33	178,071.84
Saltlands	51,752.00	6,320.87	5,109.47
TOTAL	255,376.00	314,505.75	332,751.36

artesanos who collect them from estuaries and beaches along the coast. *Artesanos* face a very low "cost of entry," in that the only equipment they need is a small hand net. This ensures a very competitive industry that supplies mariculturalists with PL shrimp at minimal cost.

Pennaeus vannamei larvae live in the open ocean, but the PL shrimp spend three to five months growing in the mangrove-lined estuaries in Ecuador. This nutrient-rich environment is ideal for fast growth and also offers protection from natural predators. The shrimp then return to sea to reproduce. The mangroves also maintain water quality and provide oxygen for many species. This *externality* (a free service provided by the mangroves) is not considered in the shrimp pricing or the destruction of the habitat.

Shrimp mariculturists originally developed ponds within intertidal salt flats; however, as the demand for shrimp increased, these flats became crowded and growers began to clear the mangrove forests (Table 22-2). Between 1979 and 1991, one-fifth of Ecuador's mangrove forests were cleared for shrimp mariculture.

During the early 1980s shrimp production in Ecuador rose by 600%. This was primarily caused by two factors. In 1982, the area experienced the *El Nino* phenomenon, which was a southward shift in warm ocean current. This increased the reproduction rate among *P. vannamei,* which created an unusually high supply of PL shrimp. Meanwhile, a large demand maintained the price for shrimp in the United States. The abundance of PL shrimp and sustained high demand induced mariculturalists to clear more and more mangrove forest to create ponds. Massive cutting of mangrove resources was possible because, until 1985, the mangroves were located on land that was considered "commons" and was unregulated by the government.[13]

Thus, market forces caused an overexpansion of shrimp pools at the expense of the mangrove forests. The destruction of the mangroves resulted in declines in the shrimp population. The result is that fewer of the ponds are being utilized, and mariculturalists are paying higher and higher prices for the diminished supply of PL shrimp, while Ecuador faces increased competition from more productive foreigners and, most importantly, a lower supply of post larval shrimp.

Table 22-2 ■ Ecuador Shrimp Ponds Expansion

Year	Hectares
1978	439
1981	34,638
1986	109,050
1991	131,961
1995	178,071

---

[12]www.earthsummitwatch.org/shrimp/national_reports/ecuador.
[13]www.earthsummitwatch.org/shrimp/national_reports.

Major problems stem from the allocation of property rights and the legal system. All mangrove ecosystems, beaches, and estuaries are considered public access for shrimp collection. This causes drastic exploitation of the PL shrimp. No attempts have been made to regulate the *artesano* suppliers. Laws were passed in 1975 and 1985 outlawing the conversion of mangrove forest to shrimping ponds; however, deforestation was reportedly continuing at a rate of 3000 hectares annually. Underpriced concessions to build shrimp ponds are given for a ten-year period, which is too short a time period to develop a sustainable focus. A long-term approach would protect habitats that enhance the stock of PL shrimp.

A potential private solution to the problem would be to create cooperative zones, where the mangroves are protected, which would lead to high availability of PL shrimp. This requires a focus on a local area, since the effects of deforestation has a local effect on the shrimp population. Secondly, agreements between mariculturalists and *artesanos* must be exclusive, ensuring that shrimp stay within the same geographical area throughout their life cycle.

Conclusion

Earth Summit Watch concluded that the shrimp industry had created an artificially high capacity to satisfy the high demand of the early 1980s at the expense of long-run sustainability. This over-investment occurred at the expense of the mangrove forests, which in turn led to a decimation of the post-larval shrimp supply.

Question 4: The author refers to an *externality* above. What is an "externality"? To what externality does the author refer?

Question 5: What is the life cycle of *Pennaeus vannamei*? How is this related to the mariculture industry?

Question 6: How is the preservation of the mangrove environment related to the viability of shrimp mariculture in Ecuador?

Question 7: What is the single most important factor the author concludes to be at the base of the financial troubles of the mariculture industry in Ecuador?

Question 8: How does the mangrove environment protect the post-larval stage of *Pennaeus vannamei*?

Question 9: What percentage of Ecuador's mangroves had been cleared by 1991? Why?

Question 10: Graph the change in mangrove area from Table 22-1. On the same graph, show the growth of shrimp ponds from Table 22-2. Interpret the trends. What additional evidence would you like to have to establish a causality between loss of mangroves and growth of shrimp ponds?

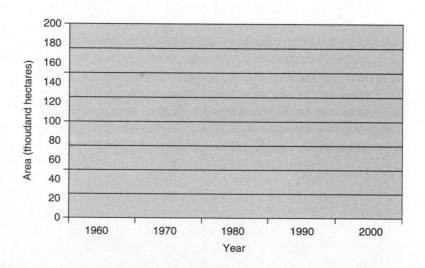

Question 11:   How did *El Nino* contribute to the rapid expansion of the mariculture business? Trace the conditions that led to the destruction of the mangroves.

Question 12:   List the two "major problems" with the system, as identified by the reviewer.

Question 13:   What solution does Earth Summit Watch offer? What basis or evidence is offered?

FOR FURTHER STUDY

- For a 1994 summary of negative impacts of coastal aquacultural development, go to http://darwin.bio.uci.edu/~sustain/shrimpecos/louis.html. This scientific article has numerous references and can provide the bases of much additional research on this subject.
- Visit the following websites and summarize the environmental costs of shrimp and salmon farming: http://www.stanford.edu/dept/news/relaged/981104shrimp.html, http://www.enn.com/enn-news-archive/1998/11/110998/aquaculture.asp.
- Shrimp aquaculture is considered one of the greatest threats to mangroves. To learn more about mangroves and their ecological significance, visit these websites: http://www.mangrove.org/sect1.htm, http://ramsar.org/about_mangroves_2.htm.
- Examine an alternative view of the impact of shrimp farming. Visit this website, http://www.agri-aqua.ait.ac.th/Mangroves/Policy.html and read the editorial. Summarize the major point of this editorial.
- Listen to this audio recording and answer the questions that follow: http://www.enn.com/enn-multimedia-archive/1999/02/022299/022299thes.asp.

  Question 1:   In what ways can the impact of salmon farming be felt thousands of miles away?

  Question 2:   How is the World Trade Organization preventing member countries from finding a solution to the farm-reared salmon problem?
- Go to http://search.npr.org/cf/cmn/cmnpd01fm.cfm?PrgDate=1%2F8%2F2001& PrgID=2 and listen to the "All Things Considered" segment on salmon farming. Summarize the major issues and analyze the viewpoints expressed in the program using critical thinking standards.

DOLPHIN-SAFE TUNA[1]

KEY QUESTIONS

- What is bycatch?
- What do consumers think "dolphin-safe" means?
- What is the exact meaning of "dolphin-safe"?
- What are the social, economic, and environmental costs and benefits of dolphin-safe fishing methods?
- Is the ecological impact of commercial fishing at acceptable levels?

One of the main problems with large-scale "harvesting" of wild marine organisms for human consumption is that most commercial fishing techniques are indiscriminate, that is, they cannot selectively capture only the target species. As a result, as much as 25% of the total global commercial catch is wasted or unused. This quantity is known as "bycatch" and refers to undersized fish, low-value, and nontarget species. These may include benthic (bottom-associated) organisms like sponges, worms, sea stars, crabs, etc., and also sharks, dolphins, whales, and sea turtles. Bycatch may die in nets or on longlines or may be returned to the water dead or dying.

Among the most harmful of all fishing activities is trawling for shrimp (see Issue 21). In addition to damaging the ocean bottom (trawling has been compared to clearcutting a forest), as much as 90% of the trawl contents may be nontarget and hence unused species, sometimes called *trash fish* by fishers, and at times including endangered sea turtles.

Question 1: Loggers sometimes refer to unwanted trees in a clearcut as "trash trees." Do you think such terminology is appropriate? Why or why not?

---

[1]Robert Young of Coastal Carolina University contributed to this Issue.

Although shrimp trawling is widespread (as many as 25,000 boats ply U.S. waters and the U.S. imports wild-caught shrimp from nearly 40 countries) and may cause extreme environmental damage, consumers are virtually unaware of the dimensions of its destructiveness. Contrasted with this is perhaps the best-publicized and galvanizing issue of bycatch—the capture of dolphins by tuna fishers.

In this issue we will analyze the multifaceted topic of bycatch in the tuna fishing industry and evaluate the costs and benefits of bycatch-reduction techniques.

DOLPHINS AND TUNA IN THE EASTERN TROPICAL PACIFIC

The eastern tropical Pacific Ocean (ETP), an area of approximately 8 million square miles (21,000,000 km^2), is one of the world's richest sources of commercially important tunas. The ETP fishery for yellowfin tuna (*Thunnus albacares*), in fact, has been called one of the most important fisheries in the world.[2] Yellowfin and skipjack tunas (*Katsuwonus pelamis*) are mainstays of the canned light meat tuna industry. The ETP fishery for albacore (*Thunnus alalunga*), whose flesh is the basis of the white-meat tuna industry, is small by comparison.

Two methods have been widely used to catch yellowfin and skipjack tunas in large-scale fisheries in the ETP. In *school fishing* (Figure 23-1), a technique no longer practiced in the ETP, rugged commercial fishers used stout rods to catch tunas, which frequently bit unbaited hooks during their feeding frenzy. Worldwide, according to Bumblebee Seafoods, 40% of the world's commercial tuna are caught on pole and line. A more productive method of catching yellowfin and skipjack tunas is *purse seining*. Globally, 30% of the world's commercial tuna are caught in purse seines. (Long-lining, in which hooks are set at intervals along a horizontal line stretching for miles, accounts for 30% of world commercial tuna catch, essentially albacore, which are also caught commercially by trolling).

In purse seining, a school of fish is encircled by speedboats with a net that may be 2 km (1.2 mi.) long and 200 meters (660 ft) deep. A purse line attached to the bottom of the net is then pulled in, trapping the tunas and other organisms unfortunate enough to be in the same location. Vessels from 12 nations, including the United States, purse seine in the ETP for tuna.

In the ETP, tunas frequently congregate around floating objects, such as tree trunks.[3] and also along with two kinds of dolphins, northern offshore spotted (*Stenella attenuata*) and eastern spinner (*Stenella longirostris*), a fact discovered by the U.S. tuna fleet nearly three decades ago. This relationship is thought to benefit the tunas, which can easily follow dolphins and take advantage of the latter's superior prey-finding abilities. Setting nets around dolphins typically catches the largest tunas and is thus the more desirable method.

When tuna seiners enter an area, they can spot aggregations of tunas and dolphins fairly easily, especially by helicopter, because dolphins are noisy and disturb the sea surface and thus are easily located. The netting process, which can take two to three hours, does not discriminate between the tunas and the dolphins, which stay together throughout the process. A number of dolphins can die during the process due directly to entanglement and drowning (Figure 23-2), and more may die later due to the delayed effects of severe trauma. It is estimated that the purse-seine fishery for tuna killed more then 1.3 million eastern spotted dolphins in the ETP between 1959 and 1990. As many as 5 million dolphins were killed during the first 14 years of purse seining in the ETP.[4]

---

[2]Joseph, J. 1994. The tuna-dolphin controversy in the Eastern Pacific Ocean: biological, economic, and political impacts. *Ocean Development and International Law 25*: 1–30.
[3]Surprisingly, enough such objects enter the ocean to be worthwhile to commercial fishers.
[4]Joseph, J. Op. cit.

FIGURE 23-1 Commercial fishers for the New England Canning Company catch tuna in the 1960s. (©Charles E. Rotkin/CORBIS)

POLICIES TO CURB DOLPHIN MORTALITY

There have been several legislative and international attempts to curb the killing of dolphins during tuna seining. The first of these was an agreement reached with the Inter-American Tropical Tuna Commission (IATTC) in 1976 (but not funded until 1979). This program sought to (1) determine dolphin mortality, (2) reduce it such that dolphin populations were not threatened and accidental killing was avoided, and (3) maintain a high level of tuna production.[5] The chief result of this effort was the placement of observers on one-third of all vessels fishing in the ETP. As a result, the first reliable estimates of dolphin mortality were made.

A further set of treaties and regulations resulted in 100% observer coverage of ETP tuna seiners and established international limits of fewer than 5000 dolphins killed by 1999. Moreover, criteria were instituted for labeling canned tuna as "dolphin-safe." As we will see, the success of the "dolphin-safe" labeling program as a deterrent to killing dolphins is unsettled. However, as a marketing tool it is unequivocal: People buy the product. For 1996, domestic canned tuna sales approached $1 billion.

There is no question that dolphin mortality has decreased in the ETP as a result of conservation measures. But the issue remains controversial and repercussions have been felt ecologically, economically, socially, and politically, as you will see below.

---

[5]Summary minutes of the 33rd meeting of the Inter-American Tropical Tuna Commission, Managua, Nicaragua, October 11–14, 1976. IATTC, La Jolla, CA., 9.

FIGURE 23-2 These photos of dead dolphins being hauled on board the Panamanian tuna boat "Maria Luisa" were from a video taken by a marine biologist who went undercover for five months on the boat to document the dolphin killings. The video from which these images were taken was broadcast on U.S. television and resulted in a huge public outcry. (AP LaserPhoto)

THE ISSUES

■ **What does "dolphin-safe" really mean?**

The Dolphin Protection Consumer Information Act (DPCIA) of 1990 established minimum criteria for tuna labeled "dolphin-safe" (Figure 23-3) in the United States. Essentially, for tuna caught from any vessel to be labeled "dolphin-safe" meant that intentional encirclement of dolphins did not occur.

A problem with this was that only about 20% of commercial tuna were caught in the ETP. There, enforcement of "dolphin-safe" capture techniques was fairly tight. However, the remaining purse-seined tuna catch was not subjected to the same stringent standards. In some cases, tuna were allowed to be designated "dolphin-safe" if the ship's skipper declared it so. Furthermore, the absence of safeguards meant that real dolphin-safe tuna, ersatz dolphin-safe tuna, and non-dolphin-safe tuna could all be found on a grocery store shelf, labeled "dolphin-safe." This also placed U.S. and other ETP purse seiners under what many considered an undue burden and hindered their ability to compete fairly.

In April 1999, then-Commerce Secretary William Daley announced that "dolphin-safe" could be used to designate any tuna harvested in the ETP if no dolphins were killed or seriously injured, even if encirclement of dolphins occurred.

FIGURE 23-3 Can of tuna labeled "Dolphin-Safe." (Louis Abel/What are **YOU** looking at Photography)

This decision was denounced by a number of nonprofit organizations. Major U.S. tuna companies announced that they would continue to honor nonencirclement policies.

On April 11, 2000, Judge Thelton Henderson of the Ninth Circuit Court of Appeals, in the case *Brower v. Daley* brought by a coalition of environmental groups and individuals, ruled that the Secretary's actions were illegal.

Judge Henderson wrote:

> (The Court) concludes . . . the Secretary acted contrary to the law and abused his discretion when he triggered a change in the dolphin safe label standard on the ground that he lacked sufficient evidence of significant adverse impacts. . . . Indeed it would flout the statutory scheme to permit the Secretary to fail to conduct mandated research, and then invoke a lack of evidence as a justification for removing a form of protection for a depleted species, particularly given that the evidence presently available to the Secretary is all suggestive of a significant adverse impact.

An alternative type of "dolphin safe" labeling has been devised by the non-profit organization Earthtrust. It awards the "Flipper Seal of Approval" to companies that meet a more stringent set of criteria.[6]

■ **Have dolphin populations in the ETP been threatened by incidental capture in tuna purse seines? Are they now?**

As we have seen, millions of dolphins have likely been killed in the ETP since the inception of dolphin encirclement. Today, that number has decreased significantly. However, according to Earth Island Institute:

> Federal scientists have determined that dolphin populations in the ETP are not recovering as expected, even with the dramatically lower reported kills of recent years. Harassment of dolphins by tuna fishermen and problems arising from the consequent physiological stress (some dolphin schools are chased and netted as often as three

---

[6]http://www.earthtrust.org/fsareq.html.

times in one day) are likely factors which cause harm to dolphin health and reproduction. Many dolphins suffer injuries in the nets and die after release, but are not counted by the on-board observer. Mothers are separated from calves, and undercounting may be occurring on board some Mexican tuna boats.[7]

To determine if a level of dolphin mortality threatens the stability of their populations, scientists must examine, among other data, the *recruitment rate* of the dolphin population. The recruitment rate is an estimate of the rate at which new individuals (i.e., recently born individuals) survive to enter the population. In this case, it provides policy makers and biologists with an estimate of the dolphin mortality that may be "acceptable," at least from the perspective of population stability. With respect to ETP dolphins, the question is whether mortalities caused by incidental catch in purse seines exceeds the recruitment rate. We will examine this issue more fully below.

■ Do alternative methods reduce dolphin bycatch?

Two methods of purse-seining in the ETP are most common—*dolphin sets* (in which nets are dropped around schools of dolphins) and *log sets* (encircling floating objects such as trees, under which fish congregate). Log sets reduce dolphin mortality, but they do so at the cost of much increased bycatch of other marine organisms.

■ What are the other issues?

Like most issues, the dolphin-tuna controversy has many dimensions. In Mexico, as many as 15,000 jobs in the tuna fishing and canning industry have been lost, and this loss has been attributed directly to the "dolphin-safe" issue.

Also, dolphins are very intelligent animals and are revered by many.

According to the National Marine Fisheries Service (NMFS) in 1975, 200,000 dolphins were killed as a result of purse-seining. The Inter-American Tropical Tuna Commission reports that about 100,000 dolphins were killed in 1989.

Question 2: What is the percent decrease in dolphin mortality over the 14-year period from 1975 to 1989?

Question 3: What is the average annual decrease in mortality over that period? (Use the formula $k = (1/t)\ln(N/N_o)$, demonstrated in *Using Math in Environmental Issues*, pages 15–16.)

---

[7]http://www.earthisland.org/news/news_immp13.html.

The estimated total population for dolphins in the ETP in 1986 was 9,576,000. Incidental kill by the purse-seine fishery was estimated at 133,174.[8]

Question 4: What percentage of the estimated dolphin population was killed by tuna seiners in 1986?

Review the section above on recruitment rate, if necessary. For dolphins in the ETP, the recruitment rate has been estimated to be about 2% of the total population.[9]

Question 5: Based on an annual recruitment rate of 2% and 1986 incidental mortality you just calculated for the ETP, do you think dolphin populations were threatened by purse seining? Show any calculations and explain you reasoning. Also, list assumptions you made in arriving at your answer.

Question 6: In light of your answer to Question 5, are efforts to reduce dolphin mortality justified from the perspective of threatening the population? Explain your reasoning.

Question 7: What do you think a typical U.S. consumer thinks upon reading the label "dolphin-safe"?

[8]Wade, P.R., & T. Gerrodette, 1993. Estimates of cetacean abundance and distribution in the eastern tropical Pacific. *Reports of the International Whaling Commission 43:* 477–493.
[9]Smith, T.D. 1983. Changes in the size of three dolphin (*Stenella spp*) populations in the eastern tropical Pacific. *Fishery Bulletin-United States, 81:* 1–13.

As previously stated, there are two common methods of purse-seining, **dolphin sets** and **log sets**.[10] A major difficulty with making tuna fishing dolphin-safe concerns the size of tuna captured by these two methods.

Dolphin sets typically kill 29 dolphins per 1,000 tons of tuna. Log sets kill less than one. Modal lengths and weights (the most commonly occurring values) of the tuna caught during log sets for 1994 were 47.5 cm and 2.1 kg for skipjack tuna. For dolphin sets, the distribution was bimodal (i.e., 2 values were equally common), 103 and 138 cm and 23 and 57 kg.[11]

Question 8: Identify a possible disadvantage to tunas of capturing such small fish using log sets. Explain your answer.

In addition to capturing immature fish (i.e., fish not yet reaching reproductive age) log sets also create another problem—bycatch. In this case, during log sets bycatch can include mahi mahi, sharks, wahoo, rainbow runners, other small fish, billfish, yellowtail, other large fish, triggerfish, and sea turtles.

For each 10,000 dolphin sets, 5340 dolphins were captured, along with 1.56 million small tunas, 11,046 sharks, 98 sea turtles, and 3641 "other small fish." The respective numbers for 10,000 log sets are 36 dolphins, 103.2 million small tunas, 140,185 sharks, 456 sea turtles, and 264,886 other small fish (Hall, unpublished data).

Question 9: For each type of bycatch above, calculate the ratio of organisms caught during log sets to those captured during dolphin sets (e.g., 36/5340 dolphins) and complete the table below.

	Log Set	Dolphin Set	Ratio
Dolphins	36	5340	1:148.3
Tunas (Small)			
Sharks			
Sea Turtles			
Other			

---

[10]Surprisingly, enough floating objects enter the ocean to be worthwhile to commercial fishers.
[11]Hall, M.A. An Ecological View of the Tuna-Dolphin Problem. Unpublished manuscript.

Question 10: For each type of bycatch, in both dolphin and log sets, calculate how many small tunas, etc. were captured for each dolphin, etc. Fill in the table below.

	Log Set	Dolphin Set
Tuna per dolphin	2.87×10^6	292
Sharks per dolphin		
Sea turtles per dolphin		
Other per dolphin		
Tuna per shark		
Tuna per turtle		

Question 11: Use the information in the tables you just constructed and your knowledge of marine life and ecosystems to answer this question: Which method—log or dolphin sets—do you think is more ecologically sound? Explain your answer.

Question 12: Recall from p. 250 the dolphin-tuna controversy in Mexico. In 1991, an embargo was placed on Mexican tuna as a result of Mexico's tuna fleet's killing too many dolphins. Do you think an embargo was justified in light of the number of jobs lost in a relatively poor country? Do you think regulations should be relaxed? Explain your reasoning.

Question 13: Even when purse seining is employed, dolphin mortality can be reduced. One method involves using finer mesh nets, which prevents dolphins from getting their snouts caught. A second, called "backing down" involves having the boat reverse after the net is set. This drops part of the net below the water line, where dolphins are herded and chased into the open sea. Fishers risk their lives by entering the water to save the dolphins. These methods have reduced dolphin mortality from 15 per set in 1986 to 3.1 in 1991 and even fewer today. Should the definition of "dolphin-safe" be expanded to include methods of dolphin sets that reduce dolphin mortality? Discuss, and justify your answer.

Question 14: Dolphins are relatively intelligent animals. What role does this play in your assessment of the issue? How do you weigh killing intelligent animals versus endangering stocks of sea turtles or sharks?

FOR FURTHER STUDY

■ **Turtle-Safe Shrimp**

The Associated Press reported:

> In early 2000, "about 280 dead sea turtles, mostly threatened loggerheads, washed up on ocean beaches on Ocracoke and Hatteras islands. Gear from large-mesh gill nets was found on four turtle bodies.
>
> The turtles were killed in greater numbers in a week's time than the average on all state beaches in a year."[12]
>
> This particular kill was attributed to commercial monkfish fishers operating in North Carolina's waters in March and April, 2000. Sea turtle kills due to commercial shrimping are even more common. In the early 1990s, according to conservationists, shrimp trawlers in Texas killed more than 11,000 sea turtles annually.[13]

Question 1: Go to http://www.enn.com/enn-news-archive/2000/04/04292000/shrimpregs_12477.asp and read the article and access linked websites. What actions have been taken and are proposed to protect endangered sea turtles? Are these sufficient to protect endangered species from extinction?

Question 2: Go to http://www.enn.com/enn-news-archive/1998/10/101398/wtoturtles.asp and http://www.sierraclub.org/trade/environment/turtles2.asp and read the article and letter. Do you think the World Trade Organization should have the right to overrule the environmental laws and regulations of the United States and other countries? Explain your reasoning.

---

[12]*Myrtle Beach Sun News,* 5/26/00.
[13]http://www.enn.com/enn-news-archive/2000/04/04292000/shrimpregs_12477.asp).

Issue 24

CORAL ROCKS! THE VALUE OF THE WORLD'S CORAL REEFS

KEY QUESTIONS

- What are coral reefs, where are they located, and what organisms contribute to reef construction and function?
- What is the importance of coral reefs as reservoirs of biodiversity?
- How important are coral reefs to marine fisheries?
- What is their significance in the carbon cycle?
- What are the global threats to the health of coral reefs?

Coral reefs are tropical, shallow-water, limestone mounds formed mainly by coral animals and plants that remove calcium carbonate ($CaCO_3$) from seawater and deposit it as skeletal material, a process known as "carbonate-fixing."

Coral refers to marine invertebrate organisms belonging to the Phylum Cnidaria, Class Anthozoa, or to the hard, calcareous structures made by these organisms. An individual coral animal, known as a *polyp,* resembles a minute *anemone.* It is the ability of these polyps to remove dissolved calcium from seawater and to deposit it as part of their rocky skeleton that allows the formation of coral reefs. How coral animals do this is not precisely known. What *is* known is that corals need algae actually living in their tissues in order to precipitate the $CaCO_3$. (More about this below.) This type of relationship, termed *symbiosis* (or more specifically, *mutualism*) occurs among a wide variety of marine organisms.

Coral reefs represent some of the most important real estate on the planet. They cover approximately 600,000 km^2 (230,000 mi^2) in the tropics, an area roughly comparable to the state of Texas. They are a major oceanic storehouse of carbon and may contain up to a million species of organisms, only a tenth of which have been identified.

Coral reefs are increasingly under threat from human actions. For example, nearly half a billion people live within 100 km (62 mi) of coral reefs (see Issues 1-3). This issue focuses on the importance of coral reefs, their role in global carbon balance, and what can be done to protect them.

BACKGROUND

While reef environments have been fairly common in Earth's history, and reefs have been built by organisms on and off for the past several billion years, not all reef structures have been built by corals. Corals have been important contributors to reefs only since about 400 million years ago, and only within the past 50 million years have modern corals assumed their reef-building roles. Noncoral reefs built by cyanobacteria were accumulating more than 3 billion years before present and thus are among the most ancient structures built by organisms.

Along with tropical rainforests, coral reefs, particularly those in the tropical Indo-Western Pacific Ocean, have the highest known biodiversities of any ecosystem on earth. But globally, coral reefs are not prospering; indeed, their very existence is being threatened by nutrient pollution, sedimentation, overfishing, global warming, and even ecotourism. The severity of this problem can be appreciated if you realize that tourists spend upwards of $100 billion each year to visit locations near reefs. Florida reefs, for example, bring in nearly $2 billion annually to that state's economy.

CORAL REEFS AND FISHING

Although reefs cover less than 0.2% of the ocean surface, they harbor a quarter of all marine fish species. Fish in coral communities are of two basic types—herbivores that feed on algae and carnivores that eat other animals.

Fishers have been plying reefs for millennia, and today reefs provide food and employment for millions. During the past few decades, however, the intensity of fishing has begun to degrade reef communities. Damage to Philippine reefs has resulted in the loss of 125,000 jobs. You may be able to anticipate some of the reasons for the degradation of coral reefs: If too many herbivores are removed, the marine algae that they eat may grow out of control and smother the coral. Removing carnivorous fish can also upset the reef's ecological balance. Can you see why?

While removal of the fish may harm reefs ecologically, some methods of collection, such as dynamiting, kill the hard coral colonies directly. Another method, called muro-ami, involves using young boys as divers who bounce rocks tethered to lines off the coral to herd the fish. This method, which originated in Japan and is employed in the Philippines, typically destroys about 17 m^2 (183 ft^2) of coral cover per hectare (10,000 m^2 = 108,000 ft^2) per operation. Typically thirty muro-ami boats repeat the process about ten times a day. It may take forty or more years for reefs that are destroyed by fishing practices to recover (if allowed to).

Macroscopic marine algae grow amid and on the coral. Some of these algae are themselves carbonate producers and may be second only to corals in their carbonate production. These algae, as well as noncalcareous algal species, may be eaten by parrotfish and other herbivores, who find safety in the numerous nooks and crannies of the reef itself. These fish in turn become food for predators.

CONDITIONS THAT FAVOR REEF GROWTH

That coral reefs are widely distributed in the tropics does not imply that the coral animals themselves are very hardy. In fact, quite the opposite is true; corals are sensitive to such a variety of environmental factors that it is a wonder they exist at all. These factors are light, temperature, salinity, sedimentation, and nutrient levels.

Light

Reef-building (also called *hermatypic*) corals are restricted to shallow waters because they require a certain quality and quantity of light. Why would simple, eyeless, invertebrate animals such as corals have a need for light? This requirement is due to the presence of one-celled algae known as zooxanthellae, which live within coral cells.

Zooxanthellae are dinoflagellates, a group that also includes the organism responsible for red tides. These organisms, thousands of which live within the cells of a polyp, require light to carry on photosynthesis. The high-energy end-products of photosynthesis are used to nourish not only the zooxanthellae but the coral as well. Some of the organic products are transferred to the coral polyp as a nutritional supplement to the food obtained when the coral feeds using its tentacles. Zooxanthellae also contribute oxygen and remove some waste materials, and they are involved in calcification.

Temperature

In addition to light, corals require a relatively constant, moderately high (but not too high) temperature. The global distribution of coral, in fact, correlates best with surface temperature: Corals are not generally found where winter surface water temperatures fall below 20°C, and they generally expel their algae and may die if water temperatures exceed 30°C. Thus, reefs are generally confined to waters between the Tropics of Cancer and Capricorn. This location protects them from low water temperatures, but means that most corals live near their upper thermal limit of tolerance.

Salinity

Salinity cannot vary much from the 35 part per thousand average for sea water for corals to survive except for Red Sea corals, which are adapted to higher salinities.

Sedimentation

Sedimentation rates should be low and grain size relatively coarse or corals can be easily smothered. First, the corals' filtering apparatus can be clogged, making it difficult for them to feed on the tiny creatures that comprise their diet, and second, the sediment can blanket the colony and keep the zooxanthellae from photosynthesizing. Sedimentation due to runoff from construction sites onshore or mining activities is one of the reefs' most lethal enemies.

Nutrient Levels

Nutrient levels (that is, the concentrations of phosphorus and nitrogen compounds) must also be low. In fact, most coral reefs flourish in nutrient "deserts." The reason again is fairly easy to understand. High nutrient concentrations may stimulate the growth of algae, which can smother the corals. Expansion of intensive, western-style agriculture on land (with its massive doses of water-soluble, nutrient-rich fertilizer) may lead to die-offs in offshore reefs. And in the United States, treated, nutrient-rich sewage from Florida's West Coast may be among sources of degradation affecting the once-healthy coral reefs in the Florida Keys.

REEFS AND THE CARBON CYCLE

Coral reefs are an integral part of the planet's carbon cycle, which may ultimately help control the earth's surface temperature range. The living polyps of a coral reef represent only a thin film covering a massive rock skeleton built up over centuries or millenia by generations of coral polyps. This hard foundation is made of calcium carbonate, which is derived from calcium ions and carbon dioxide dissolved in seawater. As the coral polyps grow, they (along with calcareous algae) precipitate calcium carbonate (that is, convert it from a dissolved form to a solid form), which supports the growing colony of coral and helps to cement coral rubble together. In the process, carbon dioxide is removed from the water. This carbon dioxide "deficit" in the water is replenished from the CO_2 in the atmosphere.

Corals are extremely efficient at fixing carbon (remember, we are talking about the carbon cycle, and carbon is a key component of calcium carbonate). Each kilogram of $CaCO_3$ contains almost 450 grams (1 lb) of carbon dioxide. (You can see why if you figure out the atomic weight of $CaCO_3$ and then determine the proportion of this that is CO_2.)

Geologists, by the way, believe that the earth's early atmosphere was greatly enriched in carbon dioxide, mainly from volcanic gases. As life evolved, some marine bacteria began to precipitate calcium carbonate, in effect storing or fixing carbon dioxide, removing it from the atmosphere. As the CO_2 was gradually removed from the early atmosphere and stored in rocks, the ability of the atmosphere to absorb heat diminished.

Today, immense volumes of carbonate sedimentary rock on continents attest to the enormous amount of carbon dioxide removed from the atmosphere. If this CO_2 were restored to the atmosphere, the earth's surface temperature might approach that of Venus, where surface temperatures (460°C) are hot enough to melt lead.

Thus, for a variety of reasons, coral reefs should be protected and their growth encouraged. Yet, in 1998, scientists documented one of the severest threats to coral survival in recorded history. The threat is called "bleaching" and it happens when corals expel the zooxanthellae living in their tissue (Figure 24-1). The greatest stress seems to come from high water temperature, which would be a consequence of global warming. Data from the past century suggest that such bleaching events could become more severe and persistent as the planet warms. This global event compounds the stresses on reefs from local human-induced sources that we described above. Exposure of coral to increased UV radiation due to a seasonal thinning of the ozone layer could be yet another significant problem.

What is the future of coral reefs? Unless humans take these threats seriously and act together to mitigate them, the future is not bright.

FIGURE 24-1 Brain coral in the Caribbean showing extensive bleaching. (© Stephen Frink/CORBIS) Corbis ID: IH089860

How Fast Are Coral Reefs Disappearing?

Consider these figures:

Radius of Earth:	6,378 km
Total area of world's oceans:	360×10^6 km^2
Total area covered by coral reefs:	>600,000 km^2
Total area of ice-free land:	1.31×10^{14} m^2

First, let's see what percentage of the planet they cover.

Question 1: We need to calculate the earth's surface. If we assume that the earth is a perfect sphere, we can calculate the total surface area using the formula $A = 4\pi r^2$. Perform this calculation below.

Question 2: If coral area is approximately 600,000 km^2, what percent of the earth's surface is covered with coral reefs?

How fast are coral reefs disappearing? One estimate, widely cited, is that 30% of the world's coral reefs will decline significantly in the next twenty years.

Question 3: Calculate the annual rate of destruction by using the compound growth formula: $k = (1/t)\ln(N/N_0)$, where k = the annual rate, t = the time in years, $\ln$ = the natural log, N = the area of reefs after twenty years, and N_0 = the area of coral reefs at the start of the twenty years. (We know the current area of coral reefs (N_0). Twenty years from now there will be 30% less coral area. You must first calculate the area of coral in twenty years before proceeding.)

Question 4: Using the annual rate of destruction for coral reefs you just calculated and a re-arrangement of the formula you just used $[t = (1/k)\ln(N/N_0)]$, determine how long it would take to destroy half of the remaining coral reefs. Repeat the calculation to determine how long it would take to destroy 99% of the world's corals. Remember to first calculate the area of coral after 50% and 99% have been destroyed. Also: Remember to use the +/- key on your calculator to enter the negative rate.

Question 5: A key assumption in your previous calculations was that the rate of destruction remains constant. Repeat the calculations in Question 4 assuming that the rate of destruction doubles.

If the rate of reef loss increases with increasing human population, the impact will be much more severe. Your calculations should have shown, even if we double the present rate of reef loss, that there is still adequate time to save this irreplaceable ecosystem, given the will to act.

BIODIVERSITY

In recent years, scientists have begun to focus on biodiversity as a central feature of ecosystem health. E.O. Wilson[1] defines biodiversity as "the variety of organisms considered at all levels, from genetic variants belonging to the same species to arrays of genera, families, and still higher taxonomic levels." One index of biodiversity is the number of species. In 1995, United Nations Environment Programme (UNEP) scientists estimated that there were currently 1.7 million named species out of a likely total of 14 million (although the range may be from 3 million to 111 million!).[2] Of these, nearly 60% are thought to be insects. UNEP projections to 2015 are that *1% to 11% of the total number of existing organisms will become extinct per decade.*

---

[1]Wilson, E.O. 1992. *The Diversity of Life* (New York: W.W. Norton).
[2]United Nations Environment Programme (UNEP) (http://www.grida.no/geo1).

Question 6: Using a computer spreadsheet or your calculator, fill in the table below, assuming an extinction rate of 0.5% per year. For the years 2010 and 2020, estimate using the compound interest formula, future value = present value * $e^{(kt)}$. In this case, the rate of growth, k, is negative since the present value is declining.

Year	Starting Number	Number Extinct	Number Remaining
1997	14,000,000	69,825	13,930,175
1998	13,930,175		
1999			
2000			
2010	XXXXXXXXX	XXXXXXXX	
2020	XXXXXXXXX	XXXXXXXX	

Question 7: With a starting number of 14 million species, use the compound growth equation again to project the number of species remaining in 50 years; in 100 years. What is the total percentage of species lost over those intervals?

Question 8: Even though the total number is declining, the doubling-time calculation introduced earlier, 70/rate, can be used here as well. What is the doubling time (or more precisely here, the halving time) for an extinction rate of 0.5% per year? In other words, when will the number of extant (i.e., living) species number 7 million?

Question 9: When will the number of species be 1/1000 of the current level? Speculate whether humans or cockroaches, rats, mosquitoes, and crows will be among the survivors. Discuss your reasoning.

Question 10:   Extinction is forever. What coral reef species would you be willing to let disappear forever? Why?

FISHING PRACTICES AND CORAL REEFS

In the United States there are approximately 565,000 marine aquarium owners and 12,000 aquarium specialty shops. To begin our analysis of the impact of removing fish from coral reefs, we need to know approximately how many fish there are currently in homes and pet stores in the United States.

Question 11:   Estimate the total number of marine fish in homes and pet stores in the United States. State the assumptions on which your estimate is based.

Even though your estimate may vary substantially from reality, it is important to at least attempt to quantify the issue. Let's see how close you actually came. According to the Pet Industry Joint Advisory Council,[3] an industry group, each of 565,000 households has 1.1 tanks, each containing 7.7 fish.

Question 12:   Assume that the cost per fish is $10.00 (a very conservative estimate for marine fish). What is the total monetary value of these fish?

The monetary value of these fish to the Filipinos who catch them is about 1% of this retail value. Estimates of the number of collectors vary, but it is likely more than 1000, and perhaps as many as 2500 full- and part-time fish and coral gatherers.[4] Despite adverse publicity and public pressure, fish capture primarily involves the use of the potent poison cyanide. As many as 90% of the tropical fish collected in the Philippines are caught by squirting them with a solution of sodium cyanide, which stuns them and facilitates their capture. Each of the 1000 or more collectors squirts about 50 coral heads each of the

---

[3]Pet Industry Joint Advisory Council. 1996. http://petsforum.com/PIJAC.
[4]Rubec, P.J. 1986. The effects of sodium cyanide on coral reefs and marine fish in the Philippines. In J.L. Maclean et al. (Eds.), *The First Asian Fisheries Forum,* (Manila, Philippines: Asian Fisheries Society).

225 days per year that reefs are fished. Although the effect of the poison on the coral polyps depends on the concentration of poison, it is known that cyanide can kill coral.

Question 13: How many coral heads are squirted with cyanide in the Philippines annually, assuming there are 1000 collectors?

Question 14: Only about 35% of the fish captured on coral reefs survive longer than 6 months in captivity, which means that 65% do not. For 7,200,000 fish surviving 6 months, how many were initially captured? How many thus died?

Question 15: Fish in coral communities are of two basic types: herbivores that crop algae and carnivores that eat other animals. Identify ways in which removal of both types of fish from reefs could adversely affect the reefs.

Although removing the fish may harm reefs ecologically, some methods of collection, such as dynamiting, kill the hard coral colonies directly. Muro-ami, which you will recall involves bouncing rocks on the coral to herd the fish, typically destroys about 17 m² of coral cover per hectare (10,000 m²) per operation. There are 30 muro-ami boats repeating the process about ten times a day.

Question 16: If the coral reefs are fished 225 days a year, how many hectares of coral will be destroyed annually?

FOR FURTHER STUDY

- Protecting Coral Reefs: Two suggestions have been made concerning protection of coral reef ecosystems, representing conflicting philosophical approaches. Some favor setting up global/regional/national reserves where economic activity is regulated, restricted, or prohibited. Other favor "privatizing" reefs, believing that owners of reefs will exert every effort to protect their investment. Assess the strengths, weaknesses, and/or limitations of each position. What additional information, if any, would you like to have to answer this question?

- Global climate change (see Issue 18) is suspected by many scientists to be directly or indirectly threatening the health of coral reefs. Identify as many potential effects of climate change as you can that could impact corals and propose reasonable means to solve these problems. Are there any that cannot be solved? Why?

- The Philippines was featured in the 1994–95 edition of World Resources[5] Analyze the following issues and how population growth affects them:
  1. Deforestation
  2. Colonization of outlying islands
  3. Expansion of muro-ami vessels
  4. Sale of tropical hardwoods
  5. Impact of fish and shrimp "farms" on mangroves

- Biodiversity of marine ecosystems is declining. Consider marine fisheries. According to the Food and Agricultural Organization of the United Nations (FAO),[6] 60% of the world's 200 major fish resources are fully fished or overfished and "need urgent management." The Atlantic Ocean was classified as "fully fished" in 1980, and the Pacific Ocean was projected to be as well by 1999. According to the U.S. National Academy of Sciences (NAS),[7] in addition to overfishing, declines in marine biodiversity are due to chemical and nutrient pollution, overdevelopment of coastal areas, introduction of exotic species (a major problem in Chesapeake and San Francisco Bays already: see Issues 19 and 20), and global climate change. Fish such as marlin, swordfish, tuna, and even sharks are now seriously threatened. Research the issue of marine biodiversity on the World Wide Web. Identify ways in which declines in biodiversity might affect ecosystems and humans. Focus on a single group of organisms (e.g., sharks, whales, kelp). Is the extinction of organisms such as sharks of concern to you? Why or why not?

- Go to http://www.state.gov/www/global/global_issues/coral_reefs/990305_coral-reef_rpt.html and read "**Coral Bleaching, Coral Mortality, and Global Climate Change**," a report released by the Bureau of Oceans and International Environmental and Scientific Affairs of the U.S. Department of State, March 5, 1999. Summarize the recommendations from the report.

- The USGS website (http://coastal.er.usgs.gov/african_dust/) states "Dust may be a viable explanation for the plight of coral reefs throughout the Caribbean." Summarize the evidence for this statement. By what mechanism is dust thought to be harming corals?

- One area where ecotourism has been exploited is the Galapagos Islands. Research the state of Galapagos reefs and describe the impacts of ecotourism on them.

[5]World Resources Institute. 1995. *World Resources 1994–95: A Guide to the Global Environment* (New York: Oxford University Press).

[6]Food and Agricultural Organization of the United Nations. 1996. Chronicles of marine fishery landings (1950–1994): trend analysis and fisheries potential. FAO Fisheries Technical Paper No. 359, by R.J.R. Grainger and S.M. Garcia. Rome.

[7]U.S. National Academy of Sciences. Understanding Marine Biodiversity (http://www.nas.edu).

- The Florida Keys is an important reef ecosystem in the United States that is threatened by indirect sources of pollution. Scientists point to sewage from Tampa-St. Petersburg and nutrients from agriculture north of the Everglades as threats to the reefs' survival. Research the issue and identify what, if any, solutions have been proposed or undertaken.

- *Reefs at Risk* is a project sponsored by the World Resources Institute (WRI), The International Center for Aquatic Resources Management (ICLARM), World Conservation Monitoring Centre (WCMC), and the United Nations Environment Programme (UNEP). Go to this website (http://www.wri.org/powerpoints/reefswww/index.htm) and view the presentation. Additional information may be obtained at http://www.wri.org/wri/indictrs/rrthreat.htm.

LEAD

KEY QUESTIONS

- How does exposure to lead affect human health?
- What are sources of lead exposure?
- What are the trends in lead emissions?
- Who is responsible for regulating lead as an environmental health hazard?

ENVIRONMENTAL HEALTH

People living in the developed West are becoming increasingly concerned about the safety of their food. In Europe, beef bans have been enacted due to fears of "Mad Cow" disease. In Belgium, dioxin-tainted foods have prompted government recalls. Some environmental groups have expressed concerns about genetically engineered foods. And throughout the United States, authorities have issued warnings about eating wild-caught fish (Table 25-1) due to contamination with one or more toxic compounds.

In June 2000, the National Academy of Sciences, the federal government's premier science advisory organization, issued a report titled "Scientific Frontiers in Developmental Toxicology."[1] Here is a portion of the Executive Summary of that report.

> Of approximately 4 million births per year in the United States, major developmental defects are identified in approximately 120,000 *live-born* (our emphasis) infants. At present, the causes of the majority of developmental defects are not understood. It is known that prenatal exposure to some chemical (e.g., lead, mercury and polychlorinated biphenyls) and physical agents (e.g., radiation) found in the environment can cause developmental defects. Scientists generally agree that approximately 3% of all developmental defects are known to be caused by exposure to toxic chemicals and physical agents, including environmental agents, and almost 25% of all developmental defects might be due to a combination of genetic and environmental factors.

It is no longer controversial that environmental toxins impose a significant cost upon humans, especially fetuses and infants. If we were to assume a minimal cost of each lost human at $1 million (based on awards in liability lawsuits), then the *minimal costs* of such toxins in the environment may exceed $3.6 billion annually in the United States alone and

---

[1]National Academy of Sciences. National Academy Press, Washington, D.C. http://www.nap.edu/catalog/9871.html and National Academy of Sciences, Washington, D.C. http://www4.nationalacademies.org/news.nsf/0a254cd9b53e0bc585256777004e74d3/85faaeadc1bcdd03852568f1006d5a5b?OpenDocument.

Table 25-1 ■ 1999 Fish Advisories
Issued by Contaminant[2]

	1998	1999
Mercury	1,931	2,073
PCBs	679	703
DDT	34	40
Dioxin	55	74
Chlordane	104	101

is likely to be much more. We will examine in the next three issues these three substances identified by the National Academy of Sciences: lead, mercury, and PCBs.

BACKGROUND ON LEAD[3]

Since 1970, emissions of lead have been cut by 98%, attributable mainly to banning lead in gasoline and in paint. Does this mean that lead is no longer a health issue?

Housing for low-income persons remains a major source of lead contamination. For example, in the District of Columbia, low-income children, many of them immigrants, are mainly housed in older buildings. Fifty-six percent of the District's housing stock was built before 1950, and more than 95 percent before 1978. Lead-based paints were used extensively before 1960, but were banned in the late 1970s. In 2000, the DC Department of Health estimated that at least 6.5% of District children had unsafe blood-lead levels, compared to the national average of 4.4 percent. Eighteen years after the District passed a strict law requiring the removal of lead-based paint from residences inhabited by children, lead reportedly continued to poison children. In 1999, the District spent over $1.5 million on lead-poisoned kids. In that year, the last year of report, at least 291 children had lead poisoning, 5 of them serious enough to need hospitalization. Each case costs around $2000 in medical costs and $4000 in special education.

The city also had problems enforcing a 1993 federal law requiring lead screening for all children entering early childhood programs. Portable lead analyzers didn't work, and three city employees assigned to be inspectors were not trained and certified to do the job. In 1999, only 29% of eligible children were screened, and of these 17,845 screenings, 2985 had missing information.

Question 1: Is it reasonable to assume that passage of federal laws ensures that enforcement will correct this problem?

---

[2]U.S. Environmental Protection Agency, Washington DC (http://www.epa.gov/ost/fish/).

[3]In addition to the sources below, information for this section was obtained from *U.S. News and World Report,* June 19, 2000. Kids at risk: Chemicals in the environment come under scrutiny as the number of childhood learning problems soar (http://www.usnews.com/usnews/issue/000619/poison.htm) and Greater Boston Physicians for Social Responsibility. May 2000. In harm's way: Toxic threats to child development, (http://www.igc.org/psr/).

Here is an example of one of the poisoned children. A 3-year-old girl living in a residence with peeling lead-based paint had blood lead-levels of 64 micrograms per deciliter, more than six times greater than the present federal standard of 10. It took three years of treatment for the child's blood level to fall to 9. During that time, the child experienced "persistent reading and writing problems."

Public schools in the District built before 1978 may themselves be sources of lead poisoning. The city estimates the cost to remediate these schools and to test all eligible children under the age of 6 to be $27.8 million over the first two years, if the project were to be approved.

Question 2: Make a preliminary assessment of a possible connection between lead poisoning and a crisis in primary and secondary education. What additional information would you like to have to address the question more fully?

Finally, some pediatricians have expressed concern that remediation of residences, if improperly done, could actually increase the risk of lead poisoning to children.

LEAD HAZARDS

The current minimum blood level that defines lead poisoning is 10 micrograms of lead per deciliter (1 dL = 100 ml = 0.1 liter) of blood. However, since poisoning may occur at lower levels than presently thought, various federal agencies are considering whether this level should be lowered further[4]

Lead has long been recognized as an extremely hazardous substance, especially to young children, fetuses, and infants; in fact, lead is one of the most toxic natural substances known, affecting virtually every system in the human body. According to the Centers for Disease Control and Prevention, lead poisoning can cause irreversible brain damage and can impair mental functioning. It may also be related to delinquent behavior among juveniles.

A recent report[5] notes that "an estimated 3 to 4 million American preschool children have blood lead levels above 10 micrograms/dl," a level now recognized to be associated with measurable neurological impairment, and as many as 68% of poor, minority children in inner cities may have unsafe lead levels.

Research implicates lead poisoning as a major risk factor in behavioral problems and criminality. One study found that lead poisoning was the strongest predictor of disciplinary problems in school, which in turn was the strongest predictor of arrests between the ages of 7 and 22. Another study, of 501 boys in Edinburgh, Scotland, found that blood lead levels correlated strongly with measures of psychological deviance.

A large-scale study by Herbert Needleman and colleagues,[6] found that children with elevated lead levels but not overt lead poisoning, suffered from reduced IQ, attention deficits, and poor school performance, and that they were seven times as likely as other children to fail to graduate from high school. Many researchers consider high lead levels

---

[4]http://emedicine.com.
[5]http://www.igc.org.
[6]http://www.crime-times.org/95c/w95cp7.htm.

to be a risk factor for crime and delinquency, because lead poisoning causes the cognitive problems most strongly linked to criminal behavior: impulsivity, low IQ, hyperactivity, and low frustration tolerance.[7]

Thus, lead can retard mental and physical development and reduce attention span. It can also retard fetal development even at extremely low concentrations. In adults, it can cause irritability, poor muscle coordination, and nerve damage to the sense organs and nerves. Lead poisoning may also cause problems with reproduction, such as a decreased sperm count. And it may increase blood pressure.

In sum, young children, fetuses, infants, and adults with high blood pressure are the most vulnerable to the effects of lead. It is estimated that approximately 930,000 children between the ages of 1 and 5 have seriously elevated blood lead levels.

Here are two reports from the Consumer Product Safety Commission:

1. "The U.S. Consumer Product Safety Commission (CPSC) and Concord Enterprises announces the recall of certain crayons imported from China because of a lead poisoning hazard. CPSC tested the crayons and found hazardous amounts of lead in the yellow and orange color crayons. If a child eats or chews on the crayon, lead poisoning could occur. Therefore, CPSC urges consumers to take the crayons away from children and discard them or return them to the store for a refund. Retailers should stop sale and return the crayons to Concord Enterprises.

The Federal Hazardous Substances Act (FHSA) bans children's products containing hazardous amounts of lead. In addition, the Labeling of Hazardous Art Materials Act amendments to the FHSA require that all art materials be reviewed by a toxicologist for chronic hazards and be labeled appropriately.

Paint and similar surface coatings containing lead have historically been the most commonly recognized sources of lead poisoning among the products within the Commission's jurisdiction. The Commission has, by regulation, banned (1) paint and other similar surface coatings that contain more than 0.06% lead ("lead-containing paint"), (2) toys and other articles intended for use by children that bear lead-containing paint, and (3) furniture articles for consumer use that bear lead-containing paint. In recent years, however, the Commission staff has identified a number of disparate products—some intended for use by children and others simply used in or around the household or in recreation—that presented a risk of lead poisoning from sources other than paint. These products included vinyl miniblinds (Figure 25-1), crayons, figurines used as game pieces, and children's jewelry.[8]

2. After testing and analyzing imported vinyl miniblinds (Figure 25-1), the U.S. Consumer Product Safety Commission (CPSC) has determined that some of these blinds can present a lead poisoning hazard for young children. Twenty-five million non-glossy, vinyl miniblinds that have lead added to stabilize the plastic in the blinds are imported each year from China, Taiwan, Mexico, and Indonesia. CPSC found that over time the plastic deteriorates from exposure to sunlight and heat, to form lead dust on the surface of the blind.

In homes where children ages 6 and younger may be present, CPSC recommends that consumers remove these vinyl miniblinds. Young children, who are most susceptible to lead, can ingest lead by wiping their hands on the blinds and then putting their hands in their mouths. Adults and families with older children generally are not at risk. CPSC found that in some blinds, the levels of lead in the dust was so high that a child ingesting dust from less than one square inch of blind a day for about 15 to 30 days could result in blood levels at or above the 10 microgram per deciliter amount CPSC considers dangerous for young children.

The Arizona and North Carolina Departments of Health first alerted CPSC to the problem of lead in vinyl miniblinds.[9]

The CPSC chided the manufacturers of these lead-containing products, saying that had they acted with prudence and foresight before introducing the products into commerce, they

---

[7]Ibid.

[8]http://www.cpsc.gov/cpscpub/prerel/prhtml94/94055.html.

[9]http://www.cpsc.gov/cpscpub/prerel/prhtml96/96150.html.

FIGURE 25-1 Photo of a recalled vinyl miniblind containing lead levels determined by the CPSC to be dangerous to children. (U.S. Consumer Product Safety Commission [www.cpsc.gov])

would not have used lead at all. This in turn would have eliminated both the risk to young children and the costs and other consequences associated with the corrective actions, which will have to be borne by other segments of U.S. society, like the healthcare system.

Question 3: Is the testing of products for toxicity a legitimate role of government? A legitimate role for the federal government?

Question 4: Would you support reducing the size and expense of government by eliminating the CPSC? Why or why not?

Question 5:   Should the monitoring of toxic substances be safely left to manufacturers or importers of suspect goods? Justify your conclusion with reasoning/evidence.

LEAD EPIDEMIOLOGY—HOW DANGEROUS IS LEAD AND WHY?

Some pediatricians have concluded that lead poisoning is the most significant chronic childhood illness related to environmental toxins.

Lead poisoning's incidence varies with the child's age and socioeconomic status, a city's population, the child's race, and—very important—the age of the home. Lead is most hazardous to the nation's 24 million children under the age of 6: In fact, one-sixth of children living in cities with over a million people, in homes built before 1946, have elevated lead levels, according to the Center for Disease Control and Prevention.

The effect of lead is complex and beyond the scope of this book; However, the best known effect is that on the production of heme. [Heme, a component of hemoglobin, is a molecule of fundamental importance in animal metabolism. Hemoglobin transports oxygen and exchanges it in the cell for carbon dioxide.] Lead interferes with two critical reactions in the formation of heme. [These reactions occur in mitochondria, structures within cells.] The result is a decrease in heme production.

This means that synthesis of a basic molecule necessary for respiration is inhibited by the presence of lead, and since this reaction occurs at the level of the mitochondria, it has a fundamental importance in human development.

The early symptoms of lead poisoning are easy to confuse with other illnesses: They may include persistent tiredness, irritability, loss of appetite, stomach discomfort, reduced attention span, insomnia, and constipation. Anemia may also be related to lead poisoning. Failure to treat children in the early stages can cause long-term or permanent health damage.

OTHER SOURCES OF LEAD EMISSIONS

Table 25-2 shows sources of lead emissions.

An "average" coal-fired power plant produces 3.5 billion kilowatt-hours in a typical year of operation. One of the by-products is lead: around 114 pounds each year.

Question 6:   How much lead is emitted for each kilowatt-hour of electricity produced? Give your answer in micrograms per kilowatt-hour.

In 1970, cars, trucks and buses alone emitted 172,000 tons of lead in their exhaust, and almost 221,000 tons were emitted from all sources. To show you how toxic that lead is, consider the federal standards for lead in the air we breathe: EPA has determined that exposure to more than 1.5 *micro*grams per cubic meter can have adverse health effects—that's only 1.5 millionths of a gram, and there are 28.5 grams in an ounce.

Table 25-2 ■ National Emissions of Lead, 1998.[10]

Source Category	Emissions (short tons[11])
Electric utilities	
Coal	54
Oil	14
Industrial fuel combustion	
Coal	13
Oil	05
Miscellaneous nonresidential fuel combustion	400
Chemical manufacturing	
Lead oxide and pigments	175
Metals Processing	
Nonferrous metals	
Primary lead	628
Secondary lead	505
Lead battery manufacture	117
Ferrous metals	
Steel production	173
Waste Disposal and Recycling	
Incineration	
Municipal waste	75
Other (hospitals, etc.)	546
Non-road Engines	
Aircraft	503

[10]*Statistical Abstract of the U.S.*, 2000.
[11]A *short ton* equals 2000 pounds.

Table 25-3 ■ Historical Data on Lead Emissions in the U.S.[12]

# of monitoring stations:		95 (for the entire U.S.!)				
Air quality standard (micrograms/m³): 1.5						
Date	**1987**	**1990**	**'93**	**'94**	**'95**	**'96**	**'97**
Air concentration of lead, based on 195 stations:	0.16	0.09	0.05	0.05	0.04	0.04	0.04
Date	**1970**	**1975**	**1980**	**1985**	**1990**	**1996**	**1998**
Total lead emitted (tons)	220,869	159,659	74,153	22,890	4,975	3,899	3,073

One of the things we need to know is the actual concentration of lead in the air. Table 25-3 contains lead emissions data.

Question 7: Plot lead emissions and air concentration on the axes below. Are the slopes of the line the same? Note the scales that we used. Try the plot with another pair of scales. Are the slopes sufficiently different to change your conclusions about the relationship?

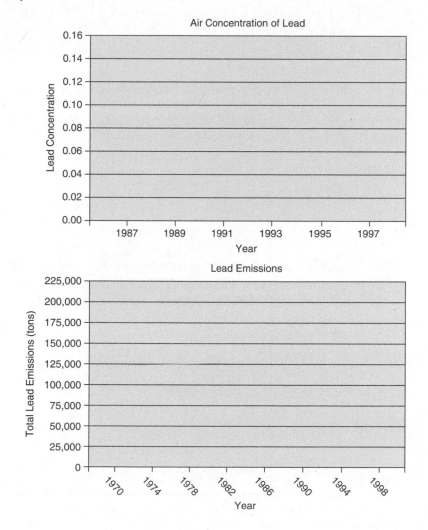

[12]http://www.census.gov/prod/99pubs/99statab/sec06.pdf.

Question 8:   Describe any relationship you observe between the atmospheric lead levels and lead emissions.

One of the objectives of the Clean Air Act of 1970 was to get that lead out of the air, and the Act targeted a major source of lead: the compound *tetraethyl lead,* which had been added to gasoline for years to boost the octane rating. Over the protests of some industry leaders, the law went into effect, and the results have been one of the nation's greatest environmental success stories. By 1975, vehicle emissions of lead were down to 130,000 tons, but the best was yet to come. By 1980, vehicular emissions were down to 60,500 tons, by 1985 to 18,000 tons, by 1990 to only 421 tons, by 1996 to 19 tons, where it has leveled off. Compare this to the amount of lead emitted from other sources in 1998: 2100 tons from metals processing, 503 tons from non-road engines and vehicles, 620 tons from waste disposal and recycling, and only 628 tons a year from primary lead production. In total, slightly under 4000 tons of lead were emitted in 1998, according to the *Statistical Abstract of the United States.*[13]

Question 9:   How many micrograms of lead are equal to 4,000 tons?

Since the area of the coterminous United States is about 8×10^{12} meters squared, there is potentially enough lead emitted annually to contaminate the lower 1000 meters of the atmosphere, *assuming* all the lead emitted was airborne and all of it remained in the air. Obviously, this is not the case. So the next question we need to ask is: Where does lead reside if not in the atmosphere?

There are only two places: the soil and the hydrosphere. Lead from old paint, vehicular emissions, incinerators and metal refining, and so on, may collect in dust, and soil, and thus is pervasive and dispersed.

SUMMARY

Let's now consider the impact of lead on animals, specifically humans. Before lead came into industrial use, human blood contained approximately 0.5 micrograms of lead in each tenth of a liter [deciliter] of blood (expressed in scientific shorthand as 0.5 micrograms per deciliter: recall that 1 microgram is a millionth of a gram and there are 28+ grams per

---

[13]Ibid.

ounce; a liter is about a quart; thus a deciliter is about half a cup.) Therefore, the "natural background level" of lead in human blood is around 0.5 micrograms per deciliter.[14]

The following information is from a report entitled "The President's Task Force on Environmental Health Risks and Safety Risks to Children" available from the Center for Disease Control's website[15] and from Rachel's *Hazardous Waste News,* which summarizes a U.S. government report.[16]

■ A child is estimated to lose two IQ points for each 10 mcg/dl increase in blood lead level.

■ Children's hearing may be impaired by blood-lead levels below 10 mcg per deciliter and hearing problems can exacerbate learning disabilities.

■ Nearly one million children living in the United States have blood lead levels high enough to impair their ability to think, concentrate, and learn. But this number has been significantly reduced from the 1980s: Then, among U.S. children 5 years old or younger, 63.3% had between 10 and 19 micrograms per deciliter; 20.5% had 20 to 29 micrograms per deciliter; 3.5% had 30 to 39 micrograms per deciliter; and 0.5% had 40 or more. Thus, the proportion of children aged 1 to 6 with lead poisoning fell to 4.4% over the period 1991 to 1994, an 80% decline from 1976–1980. But is this good enough?

Question 10: How many lead atoms would be in each liter of the blood of a child at a concentration of 10 mcg/dl?

Finally, consider this: The federal Agency for Toxic Substances and Disease Registry (ATSDR) concluded: "Lead-induced reductions in IQ not only place the individual at a disadvantage, but also eventually place the nation at a collective disadvantage in an increasingly competitive, technical, and cognition-intensive world economy." (quoted in Rachel's as cited above[17]). Clearly, lead poisoning affects society at large, and not just the individual and her or his family.

FOR FURTHER STUDY

■ In 1996 the Commission had 487 employees and a budget of around $46 million. Go to the CPSC website, www.cpsc.gov, and determine whether Americans are getting sufficient value for this expense. Justify your answer with evidence.

■ *Lead, housing and the educational system:* Citizens of Arlington County, Virginia are, like most Americans, extremely concerned about the success of their public schools. Parents and taxpayers routinely decry examples of disruptive, aggressive, hyperactive, or confrontational behavior by school children toward their peers as

---

[14]http://rachel.enviroweb.org/rhwn213.htm.
[15]http://www.cdc.gov/washington/legislative/05022000.htm.
[16]http://rachel.enviroweb.org/rhwn213.htm.
[17]Ibid.

well as teachers, but rarely is the possibility that lead poisoning may contribute to such behavior discussed (see above).

Analyze the following statistics from Arlington County, Virginia[18]

Size:	25.8 square miles
Population density	7252/square mile

Housing Stock According To Year Built

Year Built	Total Units (1990)
1989 to March 1990	2,324
1985 to 1988	5,518
1980 to 1984	3,924
1970 to 1979	7,914
1960 to 1969	15,868
1950 to 1959	18,764
1940 to 1949	19,832
Before 1940	10,703
Total	84,847

In what type of housing do you conclude the poor live—rental or owner-occupied housing? Could housing choice be contributing to behavioral problems in schools? What other information would you like to have to answer the question?

---

[18]www.census.gov, and Lead paint in apartments costs landlords $540,000. S. Chan, *Washington Post,* 2000, October 5, B5.

Issue 26

"MAD AS A HATTER": MERCURY IN THE ENVIRONMENT

KEY QUESTIONS

- Why shouldn't pregnant women eat certain kinds of fish?
- What are the sources of mercury in the environment?
- What are the health impacts of mercury pollution?
- What is biomagnification and why is it important?

JANUARY 12, 2001: FDA ANNOUNCES ADVISORY ON METHYL
MERCURY IN FISH

The Food and Drug Administration (FDA) cautioned pregnant women and women of childbearing age who may become pregnant on the hazard of consuming certain kinds of fish that may contain high levels of methyl mercury. The FDA is advising these women not to eat shark, swordfish, king mackerel, and tilefish (Figure 26-1). The FDA is also recommending that nursing mothers and young children not eat these fish as well.

Methyl mercury may harm an unborn baby's developing nervous system. These long-lived, larger fish, which feed on smaller fish, accumulate the highest levels of methyl mercury and therefore pose the greatest risk to the unborn child. Mercury can occur naturally in the environment and it can be released into the air through industrial pollution and can get into both fresh and salt water.

10 cm

FIGURE 26-1 Tilefish. (Regulatory Fish Encyclopedia, Office of Seafood and Office of Regulatory Affairs, Food and Drug Administration, 1993–2001.)

While seafood can be an important part of a balanced diet for pregnant women and those of childbearing age who may become pregnant, FDA advised these women to select a variety of other kinds of fish, including shellfish, canned fish, smaller ocean fish, or farm-raised fish. FDA concludes such women can safely eat 12 ounces per week of cooked fish. A typical serving size of fish is from 3 to 6 ounces.

As one of its priorities for fiscal year 2001, the FDA will also develop an overall public health strategy for future regulation of methyl mercury in commercial seafood.

At the same time as the FDA warnings, the federal EPA issued cautions on possible mercury contamination to women and children eating fish caught by family and friends (non-commercial fish). EPA particularly recommends that consumers check with their state or local health department for any additional advice on the safety of fish from nearby waters.

CASE STUDIES OF MERCURY POISONINGS

■ In August 1989, a previously healthy 4-year-old boy in Michigan was diagnosed with *acrodynia,* a rare manifestation of childhood mercury poisoning.[1] Symptoms and signs included leg cramps; rash; itching; excessive perspiration; rapid heartbeat; intermittent low-grade fevers; irritability; marked personality change; insomnia; headaches; hypertension; swelling; redness and peeling of the hands, feet, and nose; weakness of the pectoral and pelvic girdles; and nerve dysfunction in the lower extremities. A urine mercury level of 65 ug/L was measured on a 24-hour urine collection. Treatment with intensive chelation therapy increased his urine mercury excretion 20-fold. Examination of his mother and two siblings found urine mercury levels greater than or approximately equal to his; his father had elevated, although lower, levels.

The Michigan Department of Public Health (MDPH) identified inhalation of mercury-containing vapors from phenylmercuric acetate contained in latex paint as the probable route of mercury exposure for the family; 17 gallons of paint had been recently applied to the inside of the family's home. Samples of the paint contained 930 to 955 ppm mercury; the Environmental Protection Agency (EPA) limit for mercury as a preservative in interior paint is 300 ppm. During July, the house was air-conditioned, and the windows were not opened.

Following four months of hospitalization with repeated courses of chelation therapy and intensive rehabilitation, the patient's symptoms abated except for residual lower extremity weakness. Although certain abnormalities persist, he is able to walk and continues to improve.

■ The Texas Department of Health (TDH), New Mexico Department of Health (NMDH), and San Diego County Health Department (SDCHD) investigated three cases of mercury poisoning among persons who had used a beauty cream produced in Mexico.[2] The investigations implicated the beauty cream as the source of the mercury. The cream, marketed as "Crema de Belleza—Manning," lists "calomel" (mercurous chloride) as an ingredient and was found to contain 6 to 8% mercury by weight. In April 1996, a neurologist in El Paso, Texas, diagnosed mercury poisoning in a 35-year-old woman who resided in New Mexico; urinary mercury levels were 355 ug/g creatinine (normal: 0–25 ug/g creatinine). Beginning in September 1995, the patient had onset of symptoms progressing to *paresthesias* (left forearm, right leg, and ear), irritability, and insomnia by March 1996. A collaborative investigation by the NMDH and TDH indicated that the woman had used "Crema de

[1]http://www.cdc.gov/mmwr/preview/mmwrhtml/00001566.htm.
[2]http://www.cdc.gov/epo/mmwr/preview/mmwrhtml/00041544.htm.

Belleza—Manning" for approximately ten years and had no other known exposures to mercury. She was immediately advised to discontinue use of the cream.

Question 1: Is protecting public health from mercury poisoning a legitimate role of government? Why or why not? If not government, then who should assume this role?

Question 2: Are individuals equipped to solely assume this responsibility? Explain your answer.

MERCURY IN THE ENVIRONMENT

Mercury is widespread and persistent in the global environment, and most of its distribution is the result of human activity. In the United States, atmospheric deposition is highest in eastern Pennsylvania, Maryland, eastern Virginia, and coastal South Carolina, although there are "hot spots" elsewhere. The widespread use of mercury and its emission by combustion of fuels and waste have resulted in (1) well-documented studies of human poisonings, (2) high-level exposures to groups of industrial workers, (3) global chronic low-level environmental exposures, and (4) emissions as a result of agricultural production (mercury can be found in agricultural fungicides). Consumption of contaminated fish (especially tuna, swordfish, shark, and whale), is believed to be the major source of human exposure at the present time (see above).

Mercury in the atmosphere can be transported long distances and can contaminate isolated and distant sites. For example, there are no natural sources of mercury in the Florida Everglades, or in Minnesota's Voyageurs National Park, but mercury contamination is widespread in both places. Contamination commonly results from air deposition, but may result from groundwater contamination or surface runoff as well.[3]

Mercury is still used in many industrial activities and is found in common items such as batteries, thermometers, barometers, and even jewelry. Before 1990, paints contained mercury as an anti-mildew agent. In dentistry, mercury is still used in dental amalgams. All of these sources, and others, can release mercury into the environment.

Other than natural sources from volcanoes and mineralized veins, humans generate mercury by burning coal, by burning medical and other waste that contains mercury, by mining gold and mercury, and from industrial activity, as noted above. A typical 500-megawatt coal-burning power plant emits 170 pounds of mercury each year, according to the Union of Concerned Scientists.[4] For years, however, operators of the coal-fired power

---

[3]National Academy of Sciences. 2000. Toxicological effects of methylmercury (NAS, July 2000) Washington, DC. http://books.nap.edu/books/0309071402/html/index html.
[4]http://www.ucsusa.org/.

plants and trash incinerators responsible for substantial mercury emissions have contested attempts to further regulate mercury.

Public alarm over wildlife contamination and human health problems due to mercury (symbol Hg) toxicity has increased considerably since the mid-1960s. One issue contributing to this concern has been the number of fish consumption advisories in the majority of U.S. states, Canada, and several European countries because of high levels of mercury in game fish. Although the specific sources for this contamination are not completely known, both external sources and ecosystem-specific factors can result in toxic levels of mercury in game fish. Because mercury is known to adversely affect the human brain and nervous system (see below), health concerns arise when elevated concentrations of mercury are detected in game fish wherever there is significant human consumption (the Food and Drug Administration has no enforceable limit for mercury in fish—only a guideline of 1 part per million). Mercury concentrations in fish from the Everglades, for example, consistently exceed the Florida advisory level of 1.5 parts per million. There, even occasional fish consumption is not recommended.[5]

MERCURY IN FISH

In February 2000, the Contra Costa County (CA) Health Services Department issued an interim fish advisory for San Pablo Reservoir. Elevated levels of mercury were found in largemouth bass. Pesticides, polychlorinated biphenyls (PCBs: see Issue 27), dioxins/furans, and mercury were found in channel catfish and other species of fish in the reservoir.[6]

In April 2000, the California Office of Environmental Health Hazard prepared a draft report, "Evaluation of Potential Health Effects of Eating Fish from Black Butte Reservoir (Glenn and Tehama Counties): Guidelines for Sport Fish Consumption."[7] The report notes mercury contaminated fish from Black Butte Reservoir and provides guidelines for limiting consumption of fish taken from the Reservoir.

When toxicologist David Brown helped prepare a mercury study for eight Northeastern states and three Canadian provinces in 1997, he knew that fish in the Region's lakes would contain mercury; he just didn't know how much. As it turns out, the numbers were higher than he expected. "The most pristine lakes," he says, "had the highest levels." Brown, formerly with the Agency for Toxic Substances and Disease Registry, concluded that a pregnant woman who ate a single fish from one of these lakes could, in theory, consume enough mercury to harm her unborn child.[8]

According to EPA estimates, between 1 and 1.6 million women in the United States of childbearing years eat enough mercury-contaminated fish to risk damaging the brain development of their children. Forty states have issued one or more health advisories warning pregnant women or women of reproductive age to avoid or limit fish consumption. Ten states have issued advisories *for every lake and river within the state's borders*—Connecticut, Indiana, Maine, Massachusetts, Michigan, New Hampshire, New Jersey, North Carolina, Ohio, Vermont.[9]

---

[5]http://sofia.usgs.gov/publications/fs/166-96/.
[6]California Environmental Protection Agency http://www.calepa.ca.gov/.
[7]http://www.oehha.ca.gov/fish/special_reports/BBCRNR.html#download.
[8]"Kids at risk" June 19, 2000. *U.S. News and World Report* (cover story).
[9]Prevalence and health impact of developmental disabilities in U.S. children. *Pediatrics, 3*: 399–403, 1993; And, U.S. Census Bureau Population Estimates Program, http://www.census.gov/population/estimate/nation/inttfile2-1.txt.

MERCURY AND HUMAN HISTORY

Humans have been using mercury in some form for centuries. Mercury was an essential component of medicines, including diuretics, antibacterial agents, antiseptics, and laxatives. By 1500 BCE the Egyptians were using mercury, as it has been found in their tombs. By the end of the eighteenth century, medicines to treat syphilis contained mercury. And during the 1800s the phrase, "mad as a hatter" was coined due to the chronic mercury poisoning experienced by felters in the hat trade.[10]

MERCURY IN HUMAN MEDICINES

Due to health concerns, most medicinal uses of mercury have been eliminated and, as a result, drug-induced sources of mercury toxicity are rare in developed countries. Among the more noteworthy incidents involving Hg are human contamination in Minamata Bay in Japan (1960), methylmercury-treated grain in Iraq (1960 and 1970), and the contamination of the Everglades mentioned above. In humans, toxicologists know that mercury poisoning is commonly misdiagnosed because of the often-delayed onset of symptoms, nonspecific signs and symptoms, and lack of knowledge within the medical profession. Indeed, without a complete history, mercury toxicity, especially in the elderly, can be misdiagnosed as Parkinson's disease, senile dementia, depression, or even Alzheimer's.

In 1999 the American Academy of Pediatrics (AAP) and the U.S. Public Health Service (USPHS) issued a joint statement alerting physicians and the public to thimerosal, a preservative containing mercury used in some vaccines.[11]

Thimerosal is a derivative of ethylmercury and has been used as an additive to vaccines since the 1930s because it is effective in preventing bacterial contamination, particularly in opened multidose containers. While there was no evidence of any harm caused by low levels of thimerosal in vaccines and the risk was only theoretical, the American Academy of Pediatrics recommended in 2000 that vaccine manufacturers phase out thimerosal expeditiously. The elimination of mercury from vaccines was considered to be a reasonable means of reducing an infant's total exposure to mercury when other environmental sources of exposure are more difficult or impossible to eliminate.

MINING, ENERGY PRODUCTION, AND MERCURY

Fall 2000: The California State Water Resources Control Board, Department of Toxic Substances Control, and the U. S. Department of Forestry started a pilot program to recover mercury left over from gold mining in the Sierra Nevada foothills. The program, which began summer 2000, collected and recycled more than 200 pounds of the toxic metal. Mercury was used during the California Gold Rush to separate gold from waste. The poisonous residue ended up in river and stream banks. Until now, modern-day gold miners who found and collected the leftover mercury did not have an easy or affordable way to properly dispose of the material. Shipping it to a recycler is costly because mercury is a hazardous substance that requires a special permit to handle, store, and transport amounts greater than 10 pounds.[12]

---

[10]http://www.emedicine.com/emerg/topic813.htm.
[11]Ibid.
[12]California Environmental Protection Agency (http://www.calepa.ca.gov/).

Question 3:   Who should be responsible for ensuring that mine sites are not left contaminated? Explain your answer.

Mining for mercury at some sites has gone on for centuries, and most such sites are now contaminated. One such mine in California is an EPA Superfund site. In Slovenia, one site was mined continuously for over 500 years, only recently being abandoned. Hg wastes have contaminated a wide area in the adjacent Adriatic Sea.

In the United States, up to 1600 sites were mined for mercury to obtain the metal for use in recovery of gold from ore. The process, called "amalgamation" is a simple one. Crushed gold ore is passed over plates coated with mercury, to which the gold binds. The gold is then released from the amalgam by heating or "retorting," a process that historically released significant quantities of Hg into the atmosphere, where it ended up largely in soil and sediment. Large amounts of mercury were used, since the average concentration of gold in ores in the United States was around one-fifth ounce per ton of rock! In the Carson River Valley of Nevada, much of the sediment is presently mercury-contaminated due to gold mining.

In the developing world, especially Brazil and the former Soviet Union, gold mining and processing using mercury is generating significant quantities of mercury pollution. Fish in the Amazon River have been reported to be heavily contaminated with Hg. In Brazil, labor is cheap, and the amalgamation method is simple: Crush the ore, amalgamate it and then burn the amalgam to vaporize the mercury and release the gold. The potential for exposure is high because mercury has a high *vapor pressure,* which means that whenever ore concentrates containing mercury are heated, the mercury is vaporized. Ironically, miners have low exposure to mercury since the Hg emitted is mainly elemental Hg. However, elemental mercury may be converted to methylmercury (MeHg: see below) in soils and waters, and may thereby enter the food chain. As a result, residents who eat fish from the Amazon are exposed to a much higher risk from mercury poisoning than miners, since MeHg is more toxic than elemental Hg.

Coal normally contains mercury. Background values in coal are typically from 50 to 70 ng/g but may commonly be higher. Mercury content varies considerably from site to site, and even within coal seams. In one Kentucky coal seam, Hg content was highest at the top of the seam, and low elsewhere. Theoretically, then, the operator could have avoided much of the mercury by avoiding mining the top of the seam. Practically, however, that is often impossible. Thus, when coal is burned, mercury is volatilized. It enters the flue (exhaust) gases and does not accumulate in the coal ash.

Question 4:   A typical 500 megawatt coal plant burns about 1,430,000 tons of coal per year. At a concentration of 60 ng/g, how much mercury would such a plant emit?

NATURAL FORMS OF MERCURY

Elemental mercury is liquid at surface conditions, but it is usually bound up as minerals in nature: the most common mercury-bearing mineral is HgS, cinnabar, a distinctive red mineral. Mercury may enter the environment in the following forms[13]:

- Inorganic mercury (metallic mercury and inorganic mercury compounds) enters the air from mining ore deposits, burning coal and waste, and from manufacturing plants. It enters the water or soil from natural deposits, disposal of wastes, and possibly from volcanic activity.
- Methylmercury (CH_3Hg^+ or MeHg) may be formed in water and soil by bacteria (see below).
- Mercury in sediments and water bodies is converted by bacteria into methyl mercury.

The EPA has set the minimum risk level for oral exposure to methylmercury at 0.1 micrograms/kg/day. (Remember this number for the tuna sandwich calculation below.) Methylmercury builds up in the tissues of fish (bioaccumulates). Larger and older fish tend to have the highest levels of mercury.

HOW IS MERCURY TOXIC?

Mercury tends to bind to specific chemical structures (e.g., sulfhydryl groups) on proteins, which may explain many of its biological activities. The result is widespread changes in the function of cells, inhibition of protein synthesis, and formation of reactive oxygen compounds or radicals, which can damage DNA and disrupt cell division. In the nervous system, mercury interferes with development of small tubular structures in the neuronal skeleton. Mercury also disrupts cell membrane integrity and alters the chemical characteristics of the surface of cells, making them more likely to stick to one another. This may explain how cellular migration is altered during brain development. Mercury exposure also disrupts synaptic transmission, a basic brain function.

Mercury also exhibits synergistic[14] effects with other toxicants like polychlorinated biphenyls (PCBs). EPA reports that a study of children born to mothers who consume fish from Lake Ontario shows that prenatal PCB and mercury exposures interacted to reduce performance of 3-year-old children on certain tests. Mercury exposures in this study were quite low, yet they combined with PCB exposures to increase adverse impacts on neurodevelopment.[15]

Question 5: Discuss how mercury could be implicated in the "crisis" in our system of primary and secondary education. Cite evidence for your conclusions.

---

[13]http://www.atsdr.cdc.gov/tfacts46.html.
[14]The combined effect is greater than the sum expected by adding each.
[15]Greater Boston Physicians for Social Responsibility, *op. cit.*

Table 26-1 ■ Toxicology of Mercury (from various sources)[16]

	MeHg	Elemental Hg (Hg)	Ionic Hg (Hg^{2+})
Sources	fish and shellfish; animals fed fish meal	dental amalgams, work, fossil fuel burning, incinerators	oxidation of Hg demethylation of MeHg salts like $HgCl_2$
Usual Monitoring How it may enter your body	hair, blood	blood, urine	blood, urine
Inhalation	vapors readily absorbed	80% absorbed	aerosols of $HgCl_2$ absorbed
Oral	readily absorbed through GI* tract (eating fish)	low absorption	7–15% of $HgCl_2$ absorbed through GI
Skin	low absorption	low	moderate
Distribution in the body	throughout—strongly lipophilic	throughout—lipophilic	mainly in kidneys
*Gastrointestinal			

The nervous system of animals is clearly very sensitive to mercury. MeHg and metal vapors are most harmful, because more mercury in these forms reaches the brain, where damage can be most intense. Exposure to high levels of metallic, inorganic, or organic mercury can permanently damage the brain, kidneys, and developing fetus. Short-term exposure to high levels of metallic mercury vapors may cause effects including lung damage, nausea, vomiting, diarrhea, increases in blood pressure or heart rate, skin rashes, and eye irritation. EPA has called mercuric chloride and methyl mercury possible human carcinogens. The toxicology of the three Hg varieties is summarized in Table 26-1.

Question 6: Of the three states of mercury, which is the most readily absorbed by the gastrointestinal tract? Which is the least readily absorbed?

Question 7: Which is the most readily absorbed as a vapor? Which is the least readily absorbed?

[16]National Academy of Sciences. July 2000. Toxicological effects of methyl mercury (http://books.nap.edu/books/0309071402/html/index.html).

Question 8:   Which, then, is the most toxic form of mercury overall? Explain your answer.

The devastating effects of methylmercury on the developing human brain after excessive exposure were tragically demonstrated in large-scale poisonings in Minamata Bay, Japan, during the 1950s. Residents regularly consumed fish contaminated with methylmercury from an industrial plant's effluent in the bay. Infants born there in the late 1950s developed characteristic neurological responses that included mental retardation, disturbances of walking, speech, sucking, and swallowing, and abnormal reflexes. Mothers of affected children often showed no sign of mercury poisoning.

BIOMAGNIFICATION AND MERCURY

Mercury may become concentrated in the tissues of higher carnivores because of the tendency for Hg to *bioaccumulate* and *biomagnify*. According to the U.S. Geological Survey, in ecosystems for which atmospheric deposition of Hg is the dominant source, resulting concentrations of total mercury in water are very low, generally less than 10 nanograms per liter (ng/L). The challenge to scientists is to explain the series of processes that lead to toxic or near-toxic levels of mercury in organisms near the top of the food chain (bioaccumulation), when environmental concentrations are low.

Table 26-2 illustrates the concept of *biomagnification* as applied to mercury. Here you can see that mercury works its way up the food chain, becoming more and more concentrated in the tissues of animals that eat smaller ones. This is the reason numerous fish advisories have been issued since the 1970s.

HOW MUCH MERCURY IS IN MY TUNA SANDWICH?[18]

The EPA sets "safe" reference doses for the chemicals we are exposed to through our air, water, and food. Yet it is difficult to translate those levels, expressed in micrograms and parts per million, into information that is meaningful for our daily lives. For instance,

Table 26-2 ■ Biomagnification of Mercury[17]

egg	4,800,000 ppt
^	
adult bird	N/A
^	
large fish	690,000 ppt
^	
small fish	98,000 ppt
^	
zooplankter	14,000 ppt
^	
phytoplankter	2,000 ppt
^	
water	0.1 ppt (parts per trillion)

[17]U.S. Geological Survey.
[18]Greater Boston Physicians for Social Responsibility, op. cit.

how can you determine how much mercury you are exposed to each time you eat a tuna sandwich?

Some basic information on equivalencies and abbreviations will help you do the math so you can determine how much of a chemical you may be exposed to. The first step in determining exposure is converting the various measures into equivalent units. Use the following equivalencies:

> 1 kilogram (kg) = 2.2 pounds (lb)
> 1 pound = 16 ounces = 454 grams
> 1 ounce = 28 grams (gm)

We here repeat the following units of measure that represent tiny subdivisions of the gram (gm):

> Milligram (mg) = 1/1000 gm (thousandth)
> Microgram (μg or microgm) = 1/1,000,000 gm (millionth)
> Nanogram (ng) = 1,000,000,000 gm (billionth)
> Picogram (pg) = 1,000,000,000,000 gm (trillionth)

For example, there are 1000 milligrams in 1 gram, or 1 million micrograms in that same gram.

We are generally exposed to chemicals that are contained within another medium such as air, water, or food. In order to calculate exposure we must first calculate the concentration, or the amount of the chemical that is contained in the water we drink or the food we eat. For example, if 1 gram of fish contains, on average, 1 microgram (μg) of mercury, we would express the concentration as 1 μgm/gm. Since there are a million micrograms in a gram, another way to express this concentration is 1 part per million, or 1 ppm. The following chart outlines the equivalencies:

> gm/kg = mg/g = parts per thousand = ppt (1/1000)
> mg/kg = μgm/g = parts per million = ppm (1/1,000,000)
> μgram/kg = ng/gm = parts per billion = ppb (1/1,000,000,000)
> ng/kg = picogm/gm = parts per trillion = ppt (1/1,000,000,000,000)

Since we have determined the concentration of mercury in the tuna fish, we can determine how much mercury an individual is exposed to when eating the fish. With a few basic calculations, we can calculate the mercury exposure of a woman who consumes 7 ounces of tuna per week, given an average tuna mercury level of 0.2 ppm. Assume she does not eat any other fish or shellfish or have any other significant exposures to mercury.

First we convert the ounces into metric units:

$$7 \text{ oz.} = 196 \text{ g fish}$$

Then we multiply the amount of fish consumed/week with the concentration of mercury in the fish to determine the mercury exposure per week:

> 196 gms fish/week $\times$ 0.2 μgm mercury/gm fish = 39.2 μgm mercury/week

How much mercury is that per day? Divide by 7, since there are 7 days in a week:

> 39.2 μgm of mercury/week = 5.6 μgm of mercury/day = daily mercury exposure

Typically, we standardize exposures by dividing the total exposure by the body mass. Expressing exposure on a "per kilogram" basis allows us to compare exposures among individuals of different sizes. If we assume the woman eating the sandwich is of average weight (132 pounds or 60 kg), we divide the total exposure by 60 kilograms:

5.6 μgm/60 kg of mercury/day = 0.093 μgm/kg

We have determined that the mercury exposure of a 132 lb woman (60 kg) eating 7 ounces (196 grams) of tuna per week is 0.093 μgm/kg/day. This level of exposure is just at the limit of EPA's "safe" reference dose of 0.1 μgm/kg/day.

This calculation is based on the assumption that the woman weighs 132 lbs. What would the mercury exposure be if a 50 lb child consumed the same amount of tuna over the course of a week? The child would be exposed to approximately 0.243 μgm/kg/day of mercury.

Question 9: Do this calculation for your own weight and assume the concentrations we gave you above. How much mercury would you ingest by eating that tuna sandwich? How does this compare to the mercury standard given above?

FOCUS: MERCURY IN THE EVERGLADES

Restoration of the Florida Everglades (Figure 26-2) promises to be one of the most challenging environmental tasks to be undertaken by environmental science.

Around 5.5 million people live around the margins of the Everglades today. For decades, land around what remains of the original Everglades was converted to urban, suburban, and agricultural uses. At the same time, the U.S. Army Corps of Engineers reconfigured the natural drainage into and out of the Everglades in order to protect communities from flooding and to provide water for agriculture and urban uses. Degradation of the system inevitably followed.

Along with other contaminants, mercury levels began to rise in the sediments and the living organisms in the Everglades. Most of the mercury was probably from air deposition. Most troubling has been the rise in methylmercury (MeHg), the most toxic form. MeHg is incidentally manufactured by sulfur-reducing bacteria, which use sulfate (SOx) to power the process. Source of this sulfate is primarily from coal-burning power plants and agriculture. Comprehensive studies of the Everglades ecosystem are being undertaken by scientists at the United States Geological Survey[19]

Incinerators can put Hg into the air, where it will be deposited by precipitation. Once in the soil or water column, it can be converted to MeHg by bacteria, in the absence of oxygen and in the presence of sulfate. Both sulfate and mercury can be derived from coal-burning power plant emissions. In addition, inorganic mercury in storm runoff or soil is transformed by natural (sulfate-reducing) bacteria into methyl mercury. This usually occurs in saturated soil and, again, always occurs in the presence of sulfate and the absence of dissolved oxygen. Thus, even if coal-burning and other sources were to cease immediately, there is an ample reservoir of Hg in soils to continue the contamination process for the near future.

---

[19]Details may be found at http://sofia.usgs.gov/ and
http://sofia.usgs.gov/sfrsf/rooms/mercury/achilles_heel/cause.html.

FIGURE 26-2 South Florida. What remains of the Everglades makes up the center bottom of the peninsula. To the north are sugar cane fields (faint criss-cross pattern) "reclaimed" from the Everglades. To the east is the vast developed area (light color) sprawling westwards from Miami and Ft Lauderdale. Extensive mangrove forests line the southwest coast. (Photo Researchers, Inc.)

Why is MeHg such a concern? The USGS lists four reasons:

- Methyl mercury is rapidly taken up but only slowly eliminated from the body by fish and other aquatic organisms, so each step up in the food chain (bio)magnifies the concentration from the step below.
- Bioaccumulation factors (BAFs) of up to 10 million in largemouth bass have been reported for the Everglades.
- Fish-eating birds, otters, alligators, raccoons, and panthers can have even higher bioaccumulation factors.
- Methyl mercury in the organs and tissues causes birth defects and disorders of the brain, reproductive system, immune system, kidney, and liver at extremely low levels in food.

What is being done?

The USGS and partners are involved in a comprehensive multiyear research project to identify the nature of the mercury threat to the Everglades, what can be done about it, and how the Restoration Plan could affect mercury in soils and water bodies. One concern they are taking seriously is, how will restoring the wetlands affect methyl mercury production? Since the areas were drained, mercury has been added to the soil by airborne and waterborne contamination—from incinerators, power plants, agricultural chemicals, and even possibly sewage. At the same time, sulfate was added from power plants (acid rain) and possibly from agricultural chemicals. This mix (sulfate and mercury) is likely to result in an increase in MeHg production when the areas are wetted, as this will likely reduce oxygen content and create a habitat suitable for sulfur-reducing bacteria to thrive. As seen above, these bacteria produce methyl mercury from the raw materials.

FOR FURTHER STUDY

- Go to http://www.dep.state.fl.us/everglades/About.htm, and answer the following questions.
  1. Describe the original water-storage capacity of the Everglades.
  2. Describe the changes that began in the 1880s.
  3. How much of the original Everglades have been lost to development, and how has the Kissimmee River been altered?
  4. What was the original rationale for the alteration of the Everglades?
  5. How has Florida Bay been adversely affected?
  6. What is the source and impact of phosphorus on water quality in the Everglades?
  7. Describe the restoration plan for the Everglades, and list the separate elements of the plan.
- Research the USGS research project (at http://sofia.usgs.gov/) dealing with the potential for methyl mercury pollution in the Everglades.

ORGANOCHLORINES

KEY QUESTIONS

- What are organochlorines?
- What are their two basic chemical structures?
- Why are they toxic to life?
- What are sources of organochlorines?

THE PRECAUTIONARY PRINCIPLE REVISITED

Recall the precautionary principle from the introduction to this book. Restated, it reads, "Human society should avoid practices that have the potential to cause severe damage, even absent absolute scientific proof of harm." There is no better example of the application of the precautionary principle than to the group of chemicals produced when chlorine gas interacts with organic compounds. The compounds produced are called *organochlorines*.

Chlorine gas is a greenish, heavy, extremely reactive substance that does not occur in nature, not even in that cauldron of creation, the eruption of volcanoes. Chlorine gas readily reacts with any organic chemical it encounters, producing effective bleaches, disinfectants, insecticides and pesticides, and, incidentally, hundreds of additional by-products.

Organochlorines may be organized in two fashions: one containing structures called benzene rings (the *aromatics*) and the other (referred to as *aliphatics* from the Greek for fat; fats have the chain structure as well), which consists of chains or other linear shapes of carbon atoms. To give you an idea as to how these chemicals are named, consider PCBs, which we will discuss below. The chemical name for this "family" of compounds is "polychlorinated biphenyls." A biphenyl is made of two linked benzene rings: if chlorines are added, it becomes a polychlorinated biphenyl. Some other extremely toxic chemicals are similar in structure to PCBs. DDT for example (*di*chloro*di*phenyl-trichloroehtane) has one chlorine attached to each of two benzene rings, in turn attached to a trichloroethane. The aliphatics, since they closely mimic the structure of fats, are highly bioaccumulative; for example, hexachlorobutadiene has a bioaccumulation factor of up to 17,000.

Question 1: How many chlorines are found in hexachlorobutadiene?

Chlorination radically changes the properties of organic compounds—often increasing the stability of the compound so that it may persist in the environment for decades or centuries. This property has been of great value to industry over the past hundred years and provided us with our first "safe" refrigerant, freon. (The first generation of refrigerators used ammonia as a refrigerant, and ammonia is toxic when ingested. Freons were believed to be inert.) Furthermore, since these chemicals by and large are not naturally produced, there are few mechanisms that remove or degrade them once formed. Finally, the addition of chlorine gas to organic compounds increases their reactivity.

And perhaps most important, adding chlorine to organic chemicals increases their solubility in fats and oils. This means they can, and do, accumulate in fatty tissues of animals (*bioaccumulation*) and can be passed on from generation to generation by mother's milk. They can also become concentrated in larger animals by the simple process of eating smaller ones, a process called *biomagnification*. Thus, humans can concentrate organochlorines in their bodies by eating fish, and seabirds and marine mammals can, and do, experience the same effect.

Several hundred out of the perhaps 11,000 organochlorines produced by industry have been tested by scientists, and virtually all of them have been found to adversely affect one or more of the following processes: fetal development, brain function, functioning of the immune system, functioning of the endocrine system, and/or sperm production and development. They may be mutagenic (may facilitiate genetic mutations) and carcinogenic as well. And, finally all of these effects may occur at parts-per-trillion concentrations, which have been described as "a ratio equivalent to one drop in a train of railroad tank cars ten miles long."[1]

Question 2: Assume that the above is correct. Assess the following *ad hoc* hypothesis (an *ad hoc* hypothesis in science is one that is concocted after the fact to maintain a theory in the face of data that contradict it): "Most organochlorines are safe: it's just a coincidence that almost all the ones we have tested so far are dangerous."[2]

---

[1]Thornton, J. 2000. *Pandora's Poison* (Cambridge, MA: MIT Press).
[2]Ibid.

Question 3: Restate the quotation to be a more realistic assessment of the hazards posed by organochlorines, based on the information presented above.

Question 4: Recall Jefferson's quotation concerning corporations from *Basic Concepts and Tools,* page 14. Do you believe that corporations should be held liable for any damage that their actions cause in violation of the precautionary principle? Cite evidence to support your view, recalling that opinions uninformed by evidence are of little value in scientific inquiry.

PCBs

Polychlorinated biphenyls (PCBs) are a group of over 200 structures. Their formula is complex and their atomic weight depends on the number of relatively heavy chlorine atoms in the structure. PCBs do not exist naturally on earth; they were originally synthesized during the late nineteenth century. Because of their stability when heated, they were widely used in electrical capacitors and transformers. In the 1960s scientists began to report toxic effects on organisms exposed to PCBs, and by 1977, the manufacture of PCBs was banned in the United States, the United Kingdom, and elsewhere. By 1992, 1.2 million metric tons (2.6 billion pounds) of PCBs were believed to exist worldwide, while 370,000 metric tons (810 million pounds) were estimated to have been dispersed into the environment.

PCBs are relatively heavy molecules (average atomic weight of around 360 g/mole) and are relatively insoluble in water. Concentrations in sea water can reach 1 part per million (ppm), but PCBs typically concentrate in sediment. From there, they enter the food chain mainly through the activities of organisms called sediment- or deposit-feeders, which eat sediment, extract organic matter, and excrete the rest. Like other organochlorines, PCBs are lipophilic, and thus tend to accumulate in the fatty tissues of animals. If other animals eat the deposit-feeders, the PCBs are not metabolized, but become more concentrated in the animal's fat (biomagnified). While sea water concentration is usually below 1 ppm, concentrations exceeding 800 ppm have been measured in the tissues of marine mammals. According to the Environmental Research Foundation, this would qualify the creature for hazardous waste status!

PCBs have become widespread and serious pollutants and have contaminated most terrestrial and marine food chains. They are extremely resistant to breakdown and are known to be carcinogenic. PCBs have been linked to mass mortalities of striped dolphins in the Mediterranean, to declines in orca populations in Puget Sound, and to declines of seal populations in the Baltic.

While PCBs may threaten the entire ocean, the northwest Atlantic is believed to be the largest PCB reservoir in the world because of the amount of PCBs produced in countries that border the north Atlantic.

In terms of their specific effect on life, PCBs have been shown to cause liver cancer and harmful genetic mutations in animals. PCBs may inhibit cell division, and they have been implicated in reduction of plant growth and even mortality of plants. According to a report edited by Paul Johnston and Isabel McCrea[3] for Greenpeace UK,

> Since the rate at which organochlorines break down to harmless substances (has been) far outstripped by their rate of production, the load on the environment is growing each year. Organochlorines (including PCBs) are arguably the most damaging group of chemicals to which natural systems can be exposed.

PCBs and Orcas in Puget Sound

Even though soldiers during World War II used them for target practice, orcas have become a symbol of the Pacific Northwest.

In 1999, Dr. P.S. Ross, a research scientist with British Columbia's Institute of Ocean Sciences, took blubber samples from 47 live killer whales and found PCB concentrations from 46 ppm to over 250 ppm, up to 500 times greater than those found in humans. Ross concluded, "The levels are high enough to represent a tangible risk to these animals."

Ross compared the orca population he studied with the endangered beluga whale population of the St. Lawrence estuary of eastern North America, in which a high incidence of diseases have been linked to contaminants and which have shown evidence of reproductive impairment.

For the orcas, the PCBs are likely passed from generation to generation—an example of their persistence. PCBs are highly fat-soluble and are concentrated in mother's milk. Ross said, ". . . Calves are bathed in PCB-laden milk at a time when their organ systems are developing and they are at their most sensitive."

While PCBs have been banned in the United States for twenty years, they are still being used in many developing countries. Moreover, PCB-laden waste has been transported from industrial countries to developing countries for "disposal." (PCBs can be removed from soil or sediment by incineration, but it is expensive—too expensive for most developing nations.) Approximately 15% of PCBs reside in developing countries, mostly as a result of shipments from industrialized countries. Accordingly, Ross speculates that PCBs in the Pacific could be derived from East Asian sources and could end up concentrated in the tissues of migratory salmon, which are a prime food source for the orcas.

Ross's study, one of the most comprehensive on cetaceans (whales and their relatives) to date, was done in collaboration with the University of British Columbia, the Vancouver, B.C. Aquarium, and the Pacific Biological Station of British Columbia.

J. Cummins, in a 1988 paper in *The Ecologist,* stated that adding 15% of the remaining stock of PCBs to the ocean would result in the extinction of marine mammals.

PCBs in Sea Water

Before we ask you to calculate the concentration of PCBs in seawater, we will give you an example using gold.

Gold is dissolved in sea water at a concentration of about 0.004 parts per billion (ppb). Intuitively, you probably know that 0.004 ppb is a very small amount. But how small? How much gold is in a liter of sea water?

To calculate the number of atoms of gold in sea water, you will need the following information:

- the concentration of gold (0.004 ppb)
- the atomic weight of gold (107.9 g/mole)

---

[3]Johnston P., and I. McCrea. 1992. *The Effects of Organochlorines on Aquatic Ecosystem.* London: Greenpeace International.

- Avogadro's number (6.023×10^{23} atoms/mole)
- the density of full-strength (35 parts per thousand salinity) sea water (1.028 g/ml under "standard" conditions: 25 °C and 1 atmosphere pressure).

Step 1: The first step in solving this problem is to state an equivalency: By definition, 0.004 ppb is equivalent to 0.004 grams of gold per million kilograms ($g/10^6$ kg) of seawater, or 0.004 μgrams per kilogram (g/kg).

Step 2: The second step is to find the number of moles of gold in a mass of water (we'll use kg for convenience). To do this, multiply the concentration of gold in seawater by the factor for converting grams to moles for gold. Note that 1 kg = 10^3 g.

$$(0.004 \times 10^{-6}\,g)/10^3\,g \times 1\,mole/107.9\,g = 3.7 \times 10^{-11}\,mole/10^3\,g$$

Step 3: We next need to determine the number of atoms of gold there are in 1 kg of seawater. Avogadro's number tells you how many atoms (ions, molecules, etc.) of an element or compound there are in one mole of that element. Thus, multiply the number you just calculated by the moles-to-atoms conversion factor.

$$(3.7 \times 10^{-11}\,mole)/10^3\,g \times (6.023 \times 10^{23}\,atoms)/mole = 2.2 \times 10^{13}\,atoms/10^3\,g$$

Step 4: Thus, we now have determined the number of atoms of gold in a given mass of seawater. To calculate the number of atoms in a given volume of seawater, multiply the number of atoms of gold per 10^3 g of seawater by the density of seawater:

$$(2.2 \times 10^{13}\,atoms)/10^3\,g \times 1.028\,g/ml = 2.3 \times 10^{10}\,atoms/ml$$

Step 5: Use the ml-to-l conversion factor to convert your answer to atoms per liter.

$$2.3 \times 10^{10}\,atoms/ml \times 1000\,ml/l = 2.3 \times 10^{13}\,atoms/l$$

Now we would like you to carry out these calculations using PCBs.

Question 5: Calculate how many molecules of PCB would exist in 1 L of seawater at the Virginia standard of 0.03 ppb (ppb = μgrams per liter). Assume a molecular weight of 364 g/mole. You won't need to use the density of seawater, since the concentration of PCB is given per liter of seawater and not per kilogram.

Question 6: Determine how much the oceans can absorb without exceeding the Virginia standard. (The volume of water in the oceans is 1.37×10^{21} L). Express your answer in tonnes.

Question 7: PCBs can be removed from soil by incineration, but the process is expensive, as we mentioned. Around 15% of PCBs reside in "third-world" countries, mostly as a result of shipments of waste loaded with PCBs from developed countries. What role, if any, should the more wealthy countries play in neutralizing PCBs in third-world countries? Justify your answer by listing and defending your reasons.

FOR FURTHER STUDY

■ Some critics feel that Greenpeace scientists whom we have cited as references may be too *biased* to provide reliable data on PCBs. Research PCBs on the World Wide Web, in chemistry texts, or in other sources. Cite any differences between the treatment of these chemicals by Greenpeace and the sources you identify.

■ Organochlorines are produced as a result of the chlorination of drinking water, the "purification" of treated sewage, and the bleaching of paper pulp. Research any or all of these issues and list alternatives to the processes used that rely on chlorine. Assess the costs to changes process versus the cost that is already being borne by the environment in the form of its organochlorine load.

For example, if researching costs associated with alternative methods of bleaching paper pulp, you should consider capital cost, annualized cost, as well as the benefits alternative methods would provide to the local environment, since chlorinated pulp liquids cannot be recycled and reused by the pulp mill and must be discharged into local water bodies. Consult the websites of Environmental Defense, www.edf.org; the EPA, www.epa.gov, as well as that of the Chemical Manufacturers Association. Also see *Pandora's Poison,* by J. Thornton (MIT Press, Cambridge, MA, 2000).

■ Here are some facts about the distribution of organochlorines:

Organochlorine compounds in the tissues and fluids of the general North American population	193
Organochlorine contaminants so far discovered in the Great Lakes	83
Organochlorine by-products produced by the chlorination of drinking water and waste water	40
Organochlorine by-products of hazardous waste incinerators	31

Research these issues and conclude to what extent the hazards posed by organochlorines are offset by the value provided to society by the processes listed.

INDEX